# *Einstein and the Ether*

**Ludwik Kostro**

**Apeiron**
**Montreal**

Published by Apeiron
4405, rue St-Dominique
Montreal, Quebec H2W 2B2 Canada
http://redshift.vif.com

First Published 2000


*Canadian Cataloguing in Publication Data*

Kostro, Ludwik
Einstein and the ether


Includes bibliographical references and index.
ISBN 0-9683689-4-8


1. Ether (Space)--History--20th century. 2. General relativity
(Physics) 3. Einstein, Albert, 1879-1955. I. Title.

QC177.K68 2000     530.1 C00-901214-1

Cover design by Dominic Turgeon

# Table of Contents

# FOREWORD

The July 2000 issue of the *American Journal of Physics* contains a seven-page-long review essay on three recently published books about Einstein. The reviewer, Jeremy Bernstein, himself the author of an Einstein biography, begins his essay with the question of whether too much has been written about Einstein, and he concludes it with the statement: "We seem to discover new riches in Einstein's science each year and new oddities about his life. Has too much been written about him? Not yet."

Professor Ludwik Kostro's present study of "Einstein and the Ether" corroborates this conclusion. It challenges the widespread view that Einstein in his theory of relativity abolished, once and for all, the concept of the ether from modern physical theorising. True, in his famous 1905 seminal paper "On the Electrodynamics of Moving Bodies" Einstein declared: "The introduction of a 'luminiferous ether' will prove to be superfluous inasmuch as the view here to be developed will not require an 'absolutely stationary space' provided with special properties, nor assign a velocity-vector to a point of the empty space in which electromagnetic processes take place." But, as Kostro points out, what Einstein regarded as dispensable was the notion of an ether that has been postulated by Maxwell and his followers as well as by Poincaré to serve as the medium that explains the propagation of electromagnetic waves through space. Since such a medium defines an absolute or preferential reference system Einstein could have dismissed it not only as *superfluous* but even as *incompatible* with his special theory of relativity.

However, half a year after having constructed his general theory of relativity, Einstein readmitted the notion of an ether; for in a letter to Lorentz, a strong defender of this concept, Einstein wrote in June 1916: "...the general theory of relativity is nearer to an ether hypothesis than is the special relativity theory. But this new ether would not violate the principle of relativity, because its state, $g_{\mu\nu}$ = aether, would not be that of a rigid body in an independent state of motion, but its state of motion would be a function of position determined *via* the material processes."

For similar reasons also Hermann Weyl, in the 1919 German edition of his treatise on relativity *Raum–Zeit–Materie* suggested that, because the coefficients of the fundamental metrical tensor determine which world-points interact with another or constitute a *Wirkungszusammenhang*, the term "gravitational field" should be replaced by "ether." If furthermore we recall that in 1951 also Paul Dirac in his paper "Is there an aether?" published in *Nature*, volume 168, declared, though for quantum-field theoretical reasons, that "we can now see that we may very well have an aether, subject to quantum mechanics and conforming to relativity," it is clear that, contrary to the widely accepted view, the concept of ether is far from defunct. The generally accepted contention that the famous Michelson-Morley experiment dealt the *coup de grâce* to the ether conception had been challenged as early as December 1911 when William Magie, in an address to the American Physics Society, declared: "The principle of relativity accounts for the negative result of the Michelson and Morley experiment, but without an aether how do we account for the interference phenomena which made that experiment possible?" (*Science*, volume 35, 1912)

But what is the secret of the ether's longevity? Einstein himself provided the answer when he wrote in the book *The Evolution of Physics*, co-authored in 1951 with Leopold Infeld: "This word ether has changed its meaning many times in the development of science." Einstein was certainly right. It is probably no exaggeration to say that no other term in the vocabulary of physical theorising has suffered so many semantic changes. In fact, already at the very beginning of the history of this word, when the term *aither*, a derivation of the Sanskrit *aidh* denoting an intensely burning fire, was used in the mythopoetic language of the ancient Greeks, Homer used it as a feminine, Hesiod as a masculine noun; this difference in grammatical gender shows that different conceptions had been associated with that term. For Plato and Aristotle it denoted the supramundane fifth element which later commentators identified with the *quinta essentia* and ascribed to it theological connotations. For the Stoics it was the *pneuma*, the medium for the interactions in physical processes but also the source of life. Beginning with Descartes, followed by Newton and Boerhaave, the ether became an integral part of the mechanical philosophy and reached its climax in the different ether theories of the nineteenth century. For Oliver Lodge, for example, one of the most vociferous advocates of the ether, it served as the medium not only for the propagation of electromagnetic waves but also for the transmission of thoughts in extrasensory or telepathic phenomena. The profusion of different ether conceptions is so vast that all the books that have been written on this subject, such as those by E. M. Lémeray (1922), E. T. Whittaker (1951), K. F. Schaffner (1969), L. S. Swenson (1972), G. N. Cantor and M.J.S. Hodge (1981), present only a small part of the story.

It would be wrong to assume that the different ether conceptions evolved sequentially one from the other. O. Moon, the author of a book on Fresnel and his ether, reports that in the middle of the nineteenth century fourteen disparate ether concepts had been in use at one and the same time.

Einstein, as Kostro shows in great detail, acknowledged only three kinds of ether. But what Einstein called "ether" is no longer a rarefied material medium that permeates all space, but rather the much more abstract geometrodynamic constituent of spacetime which determines the inertio-gravitational behaviour of matter. In order to understand this point let us recall that before Einstein space and time had played the role of merely a passive background in which events take place, but Einstein's theories transformed them into active participants in the dynamics of the cosmos. We should also recall that Einstein created not only the special and general theory of relativity: during the last three decades of his life he was preoccupied, albeit without success, with establishing a third theory, a unified theory that unites the gravitational with the electromagnetic forces. With each of these three theories he associated an "ether" in the above-mentioned sense. The distinction between the three kinds of "ether" finds its mathematical expression in the different properties of the corresponding gravitational potentials $g_{\mu\nu}$ of the fundamental metric tensor: the ether of the special theory of relativity is characterised by the condition that $g_{\mu\nu} = \eta_{\mu\nu}$, where the latter is the Minkowski metric, the ether of the general theory by $g_{\mu\nu} = g_{\mu\nu}$ of the Riemann metric, and the ether of the unified theory by the fact that $g_{\mu\nu} \neq g_{\nu\mu}$.

The term "ether" is unique in the history of physics not only because of the so many different meanings in which it has been used but also because it is the only term that has been eliminated and subsequently reinstated, though with a different connotation, by one and the same physicist. Although Einstein repeatedly emphasised the difference between the two usages of this term, his homonymous application of the term was not only exploited by his opponents, like Lenard, as an argument against the theory of relativity, but was also criticised even by his staunch supporters. Max von Laue, for example, the author of the earliest textbook on relativity (1911), wrote in its fifth edition (1952) that the term "ether for the set of the $g_{\mu\nu}$ (or *Führungsfeld*) "should be avoided in order not to revive the idea of a material ether." Similarly, J. L. Synge, in his well-known treatise on the special theory declared that "it is best to avoid that dangerous word in relativity." How Einstein's enemies used this homonymy in the early twenties in order to discredit his theory has been vividly recorded by Philipp Frank in his Einstein biography, where he quotes them as having said: "For a long time efforts were made to convince us of the sensational fact that the ether had been got rid of, and now Einstein himself reintroduces it; this man is not to be taken seriously, he contradicts himself constantly."

Although Einstein's triplication of the ether has lost today its scientific actuality, his theory of relativity has altered for ever man's conceptions of space, time, and motion. Even though his ideas about the ether can no longer be upheld, they certainly deserve our attention for historical reasons at least and, in particular, for a deeper understanding of Einstein's own conception of his general theory of relativity. For it was not by accident that in 1916 Einstein reintroduced the ether which he had previously discarded. At that time he abandoned Mach's positivistic philosophy in spite of the fact that, when constructing his general theory, he had been fascinated by Mach's view that inertial resistance counteracts not as an acceleration relative to absolute space but as an acceleration with respect to the masses of the other bodies in the universe. As J. Earman, M. Friedman, J. Norton, J. Stachel and R. Torretti in their recent historical studies of Einstein's work on general relativity have made abundantly clear, Einstein's conception of the nature of spacetime is intimately connected with the now much discussed "hole argument" and the "point-coincidence argument." Although Kostro does not discuss these recent studies, his presentation of Einstein's arguments for the revival of the notion of ether, as "relative ether" or "total ether", is a useful contribution to the history of the ontological status of spacetime in modern physics. We are indebted to Professor Kostro for having devoted himself to study in such detail this facet of Einstein's work and for drawing our attention to this generally unknown chapter in the scientific biography of the man whom the periodical *Time* recently named "Person of the Century."

*Max Jammer*

# Abbreviations

| | |
|---|---|
| *AdP* | *Annalen der Physik* |
| *AHESc* | *Archive for History of Exact Sciences* |
| *AoM* | *Annals of Mathematics* |
| *Asc* | *Annals of Science* |
| *ASPN* | *Archives des sciences physiques et naturelles* |
| *EA* | *Albert Einstein Archive in the Princeton University Library* |
| *FPh* | *Forum Philosophicum* |
| *JFI* | *Journal of Franklin Institute* |
| *JR* | *Jahrbuch der Radioaktivität* |
| *MA* | *Mathematische Annalen* |
| *NRRSL* | *Notes and Records of the Royal Society of London* |
| *PF* | *Postępy fizyki (Progress in Physics)* |
| *PhB* | *Physikalische Blätter* |
| *PhR* | *Physical Review* |
| *PhZ* | *Physikalische Zeitschrift* |
| *SPAW* | *Sitzungsberichte der Preussischen Akademie der Wissenschaften* |
| *SPAW(pmK)* | *Sitzungsberichte der Preussischen Akademie der Wissenschaften, phys.- math. Klasse* |
| *VDPG* | *Verhandlungen der deutschen physikalischen Gesellschaft* |
| *VgM* | *Vierteljahresschrift für gerichtliche Medizin* |
| *VNGZ* | *Vierteljahresschrift der naturforschenden Gesellschaft, Zürich* |
| *VSNG* | *Verhandlungen der schweizerischen naturforschenden Gesellschaft* |
| *ZftP* | *Zeitschrift für technische Physik* |
| *ZMPh* | *Zeitschrift für Mathematik und Physik* |

# INTRODUCTION

In the eyes of most physicists and philosophers, Albert Einstein is credited with abolishing the concept of the ether as a medium filling space (or identified with it), which was responsible for carrying electromagnetic, gravitational and other interactions. Today this opinion is echoed in textbooks, encyclopaedias and scientific reviews. However, it does not fully reflect the historical truth, and in a sense even represents a distortion: the concept of the ether acquired new meaning in 1916 and, in the relativity theory promoted and developed by Einstein, found a new and interesting application. Einstein himself emphasises this fact in the following words:

> We may still use the word *ether*, but only to express the physical properties of space. The word *ether* has changed its meaning many times in the development of science. At the moment, it no longer stands for a medium built up of particles. Its story, by no means finished, is continued by the relativity theory.[1] **a1**

Einstein denied the existence of the ether for only 11 years—from 1905 to 1916. Thereafter, he recognised that his attitude was too radical

---

[1] A. Einstein, L. Infeld, *Physik als Abenteuer der Erkenntnis* (Leiden: A. W. Sijthoff, 1949), pp. 99-100. The quotation is taken from the English version of this book: *The Evolution of Physics* (New York: Simon and Shuster, 1961), p. 153. I have made a correction on this point, where it did not correspond to the German original. In the English version we find "some physical property of space" but in German original we have "*physikalischen Eigenschaften des Raumes.*" Therefore, I have written "the physical properties of space." In my opinion the German original better expresses Einstein's idea of the new ether. As we will see, for Einstein, space endowed with relativistic physical properties constitutes the new relativistic ether. Note that the German original and the English version were published simultaneously, in 1938, in Holland by A. W. Sijthoffs Witgenersmaatschappij, N. V. (German original), in the United States by Simon and Schuster, Inc., and in England by the Cambridge University Press (English version).

and even regretted that his works published before 1916 had so definitely and absolutely rejected the existence of the ether. This change in Einstein's attitude can be confirmed by the following two statements:

> [...] in 1905 I was of the opinion that it was no longer allowed to speak about the ether in physics. This opinion, however, was too radical, as we will see later when we discuss the general theory of relativity. It does remain allowed, as always, to introduce a medium filling all space and to assume that the electromagnetic fields (and matter as well) are its states. [...] once again "empty" space appears as endowed with physical properties, *i.e.*, no longer as physically empty, as seemed to be the case according to special relativity. One can thus say that the ether is resurrected in the general theory of relativity [...] Since in the new theory, metric facts can no longer be separated from "true" physical facts, the concepts of "space" and "ether" merge together.[2] **a2**

> It would have been more correct if I had limited myself, in my earlier publications, to emphasising only the non-existence of an ether velocity, instead of arguing the total non-existence of the ether, for I can see that with the word *ether* we say nothing else than that space has to be viewed as a carrier of physical qualities.[3] **a3**

Nevertheless, from 1905 up to the time of his death Einstein never ceased to deny the existence of the ether as understood by nineteenth century physics. Thus, when we say that Einstein removed the concept of the ether from physics we are, in a sense, correct. To be precise, however, we should say that we mean solely the ether in the sense of nineteenth century physics, since—as we will see—Einstein introduced a new, relativistic concept of the ether in 1916.

In the nineteenth century many models of the ether were constructed. All of them had one common property: in one way or another, they favoured one specific reference frame. This was particularly true of the ether concept developed by a Dutch physicist, Hendrik Antoon Lorentz. Einstein denied the existence of all ethers, as they violated his principle of relativity, according to which there is no privileged reference frame for the formulation of the laws of nature.

The ether introduced by Einstein in 1916 constitutes an ultra-referential reality and can, therefore, by no means serve as a reference frame. As a result, it does not violate the principle of relativity. Thus,

---

[2] A. Einstein, "Grundgedanken und Methoden der Relativitätstheorie in ihrer Entwicklung dargestellt" (*Morgan Manuscript*), *EA* 2070.

[3] A. Einstein, "Letter to H. A. Lorentz, 15 November 1919," *EA* 16 494.

Einstein's ether may be called a "relativistic ether." Einstein used the expression "relativistic ether" only once, in his letter to Arnold Sommerfeld in reference to Eddington's ether.[4] He never used the adjective "relativistic" when writing about his new ether. He used only the adjective "new."

The term "ether" is of Greek origin. Later assimilated by Latin, it was adopted by the European languages (*cf.* Italian *etere*, French *éther*, Polish *eter*, German *Äther*, English *aether* or *ether*). The semantic history of the term has been discussed repeatedly, particularly in the form of encyclopaedic entries or in histories of philosophy or general sciences. A history of various concepts of the ether which had appeared in modern science was given by Edmund Whittaker in his classic two-volume work *A History of the Theories of Aether and Electricity*.[5] The most recent presentation of modern hypotheses and theories of the ether, which omits the present century, is the compilation edited by C. N. Cantor and M.J.S. Hodge, *Conceptions of the Ether: Studies on History of Theories of the Ether 1740–1900*.[6]

In ancient Greece the word *ether* meant the pure, upper layers of the air that—as it was believed—filled space. The word is used in that sense by Homer. Before Aristotle, Cantor and Hodge claim that there had been no theory of the ether as such. Aristotle treated the ether as the fifth element in the composition of the Universe, in addition to earth, water, air, and fire. The ether filled and made up the supra-lunar universe. Earth, water, air, and fire were composed of primary matter and form, thus being capable of transformations from one into the others, whereas the ether was so perfect that it was not capable of being transformed, and was thus indestructible.

The ether was discussed by Stoics, by the Neoplatonist Plotinus, and by Lucretius and other ancient philosophers. Jewish, Christian and Muslim philosophers also directed their thoughts to this problem, and the ether was conceptualised in various ways, as in ancient Greece. Scholars in the Middle Ages attempted to develop Greek ideas about the ether as the medium filling space. A keen interest was shown by

---

[4] A. Einstein, "Letter to A. Sommerfeld, 28/11/1926." in: A. Einstein, *Sommerfeld Briefwechsel, Red. u Koment* p. 109, ed. A. Hermann (Basel-Stuttgart: Schwabe u. Co. Verlag, 1968).

[5] E. Whittaker, *A History of the Theories of Aether and Electricity*, vol. 1-2 (New York: Harper and Brother, 1953).

[6] G. N. Cantor and M.J.S. Hodge (eds.), *Conceptions of Ether* (Cambridge: Cambridge University Press, 1981).

Renaissance scholars in ancient Greek culture and concepts of the ether. Thus, the concept of the ether entered modern sciences, playing an important role for Descartes, who never recognised the existence of a void. His "subtle matter" that filled space was called by the "Cartesian ether" by his contemporaries.[7]

Isaac Newton, who left us an outstanding legacy in the form of the mechanics (nowadays referred to as the *classical*, modern theory of gravitation), as well as important discoveries in the field of optics, used the ether to explain the transmission of gravitational interactions and various optical phenomena. Newton constructed various models and versions of the ether, though he was never satisfied with them. The idea of the ether was always present in his mind and works, as he could never believe that gravitational interactions might be transmitted by a void.[8] Newton also created the first corpuscular-wave theory of light, in which the vibrations of particles of light created waves in the ether. This theory was used by Newton to explain many optical phenomena.[9] Newton was not alone in viewing the ether as necessary for transmitting the gravitational interaction; G.W. Leibniz was of the same opinion, as shown by M. Jammer.[10]

When the phenomena of diffraction and interference of light were discovered—thanks to experiments performed by Augustin Fresnel and Thomas Young—it became obvious that light had to be treated as a wave process. The wave theory of light reinforced belief in the existence of the ether. It became the medium of transmission for light, and later, due to discoveries and work by Michael Faraday, James Clerk Maxwell, and Heinrich Rudolf Hertz, it became the transmitting agent for electromagnetic waves in general, including light. As the wave theory of light gained strength in the nineteenth century, various hypotheses, theories, models and versions of the ether emerged. These new ethers emerged particularly at the end of the nineteenth century, as the ether was identified with the concept of absolute space in classical physics. This new ether concept was promoted by Paul Drude, whose books

---

[7] *Ibid.*, pp. 8-12.

[8] I. Newton, *Mathematical Principles* (Cambridge: Cambridge University Press 1934), p. 634.

[9] J. L. Haves, "Newton's Revival of the Aether Hypothesis and the Explanation of Gravitational Attraction," *NRRSL*, 23 (1968), pp. 200-212; L. Rosenfeld. Newton's Views on Aether and Gravitation," *AHESc*, 6 (1969), pp. 29-37; E. J. Aiton, "Newton's Aether-Stream Hypothesis and the Inverse Square Law of Gravitation," *ASc*, 25 (1969), pp. 225-260; H. Guerlac, "Newton's Optical Aether," *NRRSL*, 22 (1967), 45-57.

[10] M. Jammer, *Concepts of Space* (1954; rpt. New York: Dover, 1993).

Einstein enjoyed reading. It was also promoted by M. Abraham and others.[11] There were ethers that formed a continuous medium, and there were ethers of a corpuscular nature. In some models the ether could move in space, be subject to fluctuations, or be carried or trapped by celestial objects. In other conceptions, however, the ether remained stationary, sometimes interpreted as a state of absolute rest. Some models treated the ether as a liquid, while others treated it as like a solid, particularly when it was discovered that light was made up of transverse waves. In the last century, scientists were convinced that transverse waves can propagate only in a solid environment, not a liquid milieu.

The young Einstein, who attended secondary school at the end of the 19th century, believed in the existence of an elastic ether in which light was propagated with varying velocities. As a student at the Technical University of Zurich, Einstein was at first convinced of the existence of the ether. He devised various kinds of apparatus to detect the Earth's movement through the ether. By the completion of his studies, however, he doubted the existence of any medium that propagated light. The ideas Einstein had formed about the ether by the year 1905, *i.e.*, by the time his Special Theory of Relativity was published, are presented in Chapter 1 of this work.

When Einstein published his Special Theory of Relativity he was interested only in Lorentz's concept of the ether, as he found all other concepts inferior. Lorentz's ether, which remained at perfect rest, particularly favoured one reference system—the privileged reference system in Lorentz's theory of electromagnetism. Within this system, light had its own basic constant velocity. Einstein could not agree with this concept of electromagnetism, as it introduced into physics—so Einstein claimed—an intolerable asymmetry. On the one hand, all inertial systems were equal for formulating laws of mechanics, while on the other hand, one of the inertial reference systems was privileged in formulating the laws of electromagnetism. This was the system of the ether at rest. The Special Theory of Relativity removed this asymmetry by introducing the relativity principle, yet it also led to the conclusion that Lorentz's ether was superfluous. Thus, at the beginning of his academic career, Einstein absolutely rejected the existence of the ether. This rejection, as mentioned above, lasted until 1916, when the new concept of the ether was introduced, which happened immediately after the final formulation and publication of the General Theory of Relativity. The history of the

---

[11] *Ibid.*, p. 125.

General Theory of Relativity started in 1907. It led to a gradual discovery of the real attributes of physical space (closely related to time in relativity theory), which Einstein had failed to perceive before 1916. This is how the new concept of the ether was born, even though Einstein was unaware this was happening. The period of Einstein's absolute rejection of the ether and his subconscious creation of a new concept is described in Chapter 2.

The correspondence on the General Theory of Relativity with Lorentz (who persistently defended his concept of the ether) and polemics with a German physicist, Philipp Lenard (an anti-relativist and ardent supporter of the old ether) made Einstein aware that, in his General Theory of Relativity, he had discovered real attributes of the space-time continuum, defined by the components of the metric tensor $g_{\mu\nu}$, which in his new theory were conceptualised as gravitational potentials. From that moment on, he attributed the name "ether" to the reality that is described locally by the metric tensor. The period during which the new concept of the ether was introduced and developed, which extended from 1916 to 1924, is described in Chapter 3.

The further history of the new ether is connected with Einstein's attempts to formulate a Unified Relativistic Field Theory. The four-dimensional continuum of General Relativistic space-time, or more precisely, the $g_{\mu\nu}$-field, did not include electromagnetic fields in its structure, since it was only a gravitational field. Thus, the ether of the General Theory of Relativity turned out to be a gravitational ether only. From 1929 until the end of his life in 1955, Einstein tried to build electromagnetic fields into the structure of the space-time continuum in order to unify the gravitational and electromagnetic interactions. On occasion, he was convinced that he had managed to find a satisfactory solution to the problem of unification. This happened, for instance, when he enriched Riemann's geometry with so-called tele-parallelism, and used the geometry developed in this way to search for unification. During this time, Einstein produced a veritable flood of lectures and research on the issues of space, the ether, and field, where the ether—identified with a space characterised by metrics and tele-parallelism—was understood as a medium for transmitting both gravitational and electromagnetic interactions. The years 1925-1955 are described in Chapter 4.

On the basis of Einstein's works on the new ether, we can distinguish three models of the relativistic ether (the model of the Special Theory of Relativity, of the General Theory of Relativity, and of the

Unified Field Theory with its variations) and grasp certain characteristic features of each. These models and their features are presented in Chapter 5, along with a summary of the preceding chapters. Our purpose in presenting Einstein's models of the ether and its characteristic features is to grasp its physical meaning.

According to information obtained from Professor Don Howard, one of the chief editors of documentation related to Einstein's life and research, *The Collected Papers of Albert Einstein* (Boston University, USA),[12] to date no one has presented the history of Einstein's concept of the ether. Thus, this work is the first comprehensive history of Einstein's concept of the ether. There do exist a number of articles outlining the history of this subject by the author of the present work.[13] In works by other historians of physics the author has been able to obtain, Einstein's ether and its features are given a mere mention.[14] Many documents presented or quoted in this work have never been published. The documentation I have drawn upon here has been collected by the library of the Museum of Science and Technology in Munich (*Deutsches Museum*) and in the Bayerische Staatsbibliothek in Munich. I was given the opportunity to secure the unpublished documents through the kindness of the Editor in Chief of the above-mentioned documents, Professor John Stachel, as well as Professor Don Howard. Copies of documents

---

[12] 795 Commonwealth Avenue, Boston, MA 02215 (sponsored by the Hebrew University of Jerusalem and Princeton University Press).

[13] L. Kostro, "Einstein's new conception of the ether," in: *Physical Interpretations of Relativity Theory*, ed. M. C. Duffy (London: Imperial College, 1988), pp. 55-64; "Outline of the history of Einstein's relativistic ether conception," in: *Studies in the History of General Relativity*, *"Einstein Studies,"* vol. 3, eds. Jean Eisenstaedt and A. J. Kox (Boston: Birkhauser, 1992), pp. 260-280; "The Evolution of Einstein's Ideas Concerning Ether, Space and Time," in: *History of Physics in Europe in 19th and 20th Centuries*, 1st EPS Conference on History in Europe in the 19th and 20th Centuries, Como, 2-3 September, 1992, ed. F. Bevilaqua, SIF, Conference Proceedings, Vol. 42 (Bologna: Editrice Compositori, 1993), pp. 177-183; "Albert Einstein and the Theory of the Ether," in: *Foundations of Mathematics & Physics*, eds. U. Bartocci and J. P. Wesley (Blumberg, Germany: Benjamin Wesley, 1990), pp. 137-162; "The Physical Meaning of Albert Einstein's Ether Conception," in: *Frontiers of Fundamental Physics*, eds. M. Barone and F. Selleri (New York-London: Plenum, 1994), pp. 193-201.

[14] R. Wahsner, "Äther und Materie—von Descartes bis Einstein," *Wissenschaft und Fortschritt*, 29, (1979) pp. 54-57; A. Miller, *Imagery in Scientific Thought Creating 20th Century Physics* (Boston-Stuttgart: Birkhäuser, 1984), pp. 55-60; J. Illy, "Einstein Teaches Lorentz, Lorentz teaches Einstein. Their Collaboration in General Relativity, 1913-1920," *AHESc*, 39, (1989) pp. 247-289; by the same author: "Mach, Lorentz and Einstein on Ether" in: *Ernst Mach and the Development of Physics*, Intern. Conf., Sept. 14-16, 1988, Prague.

together with the kind permission to publish them were received from the New-York-resident representative of the Jewish University at Jerusalem, Mr Benamy Ehud.

The documentation collected is by no means complete, yet—the author believes—it is sufficient to reconstruct the history of the concept presented in this book.

Einstein's works referring to the new ether are of a purely interpretative nature, and therefore they do not contain any mathematical formulae. Moreover, they lack any novel mathematical ideas which could be used to master the formalism of the theory. They serve simply to uncover the physical meaning hidden under—or perhaps behind—the mathematical apparatus of the Special Theory of Relativity, the General Theory of Relativity, and the attempts to formulate the Unified Field Theory. Einstein was skilled in utilising the mathematical apparatus, and highly talented at presenting the physical content expressed by the mathematical formalism of his theory. He was trying not to allow the mathematical apparatus (which he highly appreciated, particularly when he used it in formulating his General Theory of Relativity) to obscure the physical content. The physical meaning was something very fundamental to him; he believed that this attitude should be adopted by every physicist, and therefore praised this ability among his colleagues whenever he encountered it. He wrote to his friend, the renowned physicist Paul Ehrenfest:

> You are one of the few theoreticians who have not been deprived of their native intelligence by the mathematical epidemic.[15]

After Ehrenfest's death, Einstein wrote:

> His stature lay in his unusually well developed faculty to grasp the essence of a theoretical notion, to strip a theory of its mathematical accoutrements until the simple basic idea emerged with clarity.[16]

This same method was adopted by Einstein in his works on the relativistic ether to lay bare and explain the fundamental idea in his concept. The method is discussed in many historical parts of this book (*i.e.*, the first four Chapters), as we will frequently quote excerpts from his works and letters. The author will also follow the same method in

---

[15] A. Einstein, "Letter to P. Ehrenfest (1912)," quoted in: M. J. Klein, *Paul Ehrenfest* (Amsterdam: North Holland, 1972), p. 190.

[16] A. Einstein, "Paul Ehrenfest in Memoriam," [in:] A. Einstein, *Out of My Later Years* (New York: Philosophical Library, 1950), p. 237.

Chapter 5 of this book, where an attempt is made to present the physical meaning of Einstein's relativistic ether.

It must, however, be pointed out that in using the method he praised so highly, Einstein by no means wished to diminish the role of mathematics in the Theory of Relativity. What would his General Theory of Relativity be without the pseudo-Riemannian four-dimensional geometry? This geometry is not merely a tool or mathematical apparatus, but it is also an integral part of the theory. Only when the General Theory of Relativity is equipped with its mathematical formalism does it become a true physical theory, *i.e.*, a powerful tool for research into physical reality. By seeking to disclose the physical meaning in his works, Einstein sought to point out that, whereas good specialists may be able to manipulate mathematical formulae with ease, they often create the illusion of a profound understanding of the physical content, which is not always true. One can use the mathematical formalism of a given theory expertly without understanding its deeper physical meaning.

In rejecting the nineteenth century ether and introducing a new, relativistic ether, Einstein was motivated not only by purely physical arguments, but also by philosophical principles. At first he followed the philosophical vogue of his youth, and then he became a philosopher himself. Indeed, we may speak of Einstein's philosophy of physics and his philosophy of Nature. At first he had been influenced by the positivistic physical cognitive theory of the representatives of the so called "second positivism" (Ernst Mach, Friedrich Wilhelm Ostwald, and Richard Avenarius), but later abandoned this theory to create his own philosophy of physics, anticipating the ideas of future philosophers of science, such as Karl Popper. Consequently, the philosophical conditioning of Einstein's ideas about the ether must also be discussed here.

In this book, the reader will find many excerpts from works and letters written by Einstein, as well as other documents. Some of these are even quoted in full. By quoting long excerpts, or even whole documents, I seek to give the reader direct access to Einstein's ideas and concepts, particularly when the documents are not readily available, owing to the limited circulation of many of the journals that published some of these works, as well as the language barrier the reader may face.

The texts by Einstein and others quoted in this book can be found in the Appendix in the original language, if they were not already in

English. The numbers assigned to the texts refer to the selected sources listed at the end of the book.

This book contains very few equations, and all mathematical formulae are presented in their original Einsteinian version. Only the final chapter may occasionally reflect contemporary mathematical notation.

This book is written for physicists who want to learn about Einstein's ideas concerning the ether, as well as philosophers who deal with issues relating to space and time.

I would like to express my gratitude to all the people and institutions whose kindness made it possible for me to write this book. Permission to reprint published and hitherto unpublished works and letters by Einstein was granted by The Albert Einstein Archives, The Jewish National & University Library, The Hebrew University of Jerusalem, Israel. I am particularly grateful to the Foundation of the *Volksvagenwerk* Company in Hannover for granting me a three-month scholarship at the suggestion of Kerschensteiner Kolleg at the Deutsches Museum in Munich, which enabled me to collect most of the material used in this book; Professor John Stachel and Professor Don Howard for their kind permission to inspect unpublished works and letters by Einstein, which facilitated my writing this book. I am particularly grateful to Professor Don Howard for long discussions during our stay at Deutsches Museum, which also helped me greatly. I am also indebted to Professor Peter Bergmann, Einstein's co-worker, for conversations full of interesting information; to Professor Michal Heller, who edited the book, for his invaluable comments, which helped to avoid errors in this dissertation; Professor Franco Selleri, whose knowledge of physics and teaching talent I admire and to whom the English and Italian translations of this book owe their existence; to Professor Bronisław Średniawa, one of the leading Polish physicists and historians of physics, whose books, articles and advice have aided me considerably; and to Jill Corner and Roy Keys for editing the text. Finally, I owe a special debt of thanks to Professor Max Jammer for kindly consenting to write a foreword to this book.

# Chapter 1

# EINSTEIN'S VIEWS ON THE ETHER BEFORE 1905

## 1.1 First notions of electromagnetism and the ether

Albert Einstein was born on March 14, 1879 in Ulm, Germany. His parents, Hermann Einstein and Pauline, née Koch, lived at 135 Bahnhofstrasse. Shortly after Albert's birth, Hermann's enterprising and energetic younger brother Jacob, an engineer, proposed to Albert's father that they start a small gas and water installation business together in Munich. Einstein's father agreed and Albert's family moved to Munich. The modest undertaking had a promising beginning, but Albert's uncle had greater ambitions. As a result, a few years later, they opened an electromechanical factory to produce dynamos, arc lamps, and electrical measuring equipment for municipal electric power stations and lighting systems. As a result, Albert Einstein spent his earliest years in the capital city of Bavaria in a stable, comfortable milieu at Adelreiterstrasse.[17]

It is quite understandable that one should acquire one's first and most fundamental ideas from one's family and school environments. By the end of the nineteenth century, there was a general belief in the existence of the ether, conceived as a carrier of electromagnetic radiation. We cannot tell exactly when and where young Einstein heard about the

---

[17] A. Pais, *Subtle is the Lord..., The Science and the Life of Albert Einstein* (Oxford: Oxford University Press, 1982), pp. 35 and 37.

ether for the first time and came to believe in its existence. We do know, however, that his interest in electromagnetism was aroused in his childhood, for he was interested in what was happening at the electromechanical factory owned by his uncle and father. Through the answers he received to the questions he asked them, he gained his first information on electromagnetism and electrical engineering. In his "Autobiographical Notes" he mentions how the operation of a compass his father showed him captured his infant imagination.[18]

Young Einstein was probably interested in the practical issues solved at his uncle's and father's factory, as well as the theoretical questions. From age ten to fourteen, he read a large number of popular science books delivered to him by a poor medical student named Max Talmud, who came to dine with the Einsteins every Thursday. Albert discussed various scientific and philosophical problems with Max for hours. It is quite probable that Einstein acquired some knowledge about electromagnetism and the ether from the books he read at that time. Some of the concepts he must also have learned at school later on.[19]

In Munich, Einstein first went to the Volksschule, and then to Luitpold Gymnasium. When he was in the senior class of the school, his parents moved to Milan, Italy, as their business had started to incur losses in Germany and they expected better results in Italy where they had been exporting their products previously. Albert was left behind in Munich to complete his education at the Luitpold Gymnasium. He did not like the atmosphere and teaching style at this school, so he left Munich without letting his parents know, and moved to Pavia, Italy, where his parents were living at 11 Foscoli Street. In Pavia, the factory owned by Einstein-Garrone partnership began successfully, though it did not prosper for long.[20]

## 1.2    Einstein's youthful "scientific work" on the ether and magnetic field

In Pavia, Einstein's interest in electromagnetism and the ether bore fruit in the form of his early "scientific paper."[21] The teenage Einstein (fifteen or sixteen years old at the time) felt an urgent need to set down

---

[18] A. Einstein, *Albert Einstein: Philosopher–Scientist*, ed. P.A. Schilpp (Evanston, Ill.: Library of Living Philosophers, 1949), p. 19.

[19] A. Pais, *op. cit.*, p. 37.

[20] *Ibid.*, p. 39.

[21] J. Mehra, "Albert Einsteins erste wissenschaftliche Arbeit," *PhB*, 27 (1971) pp. 385-389.

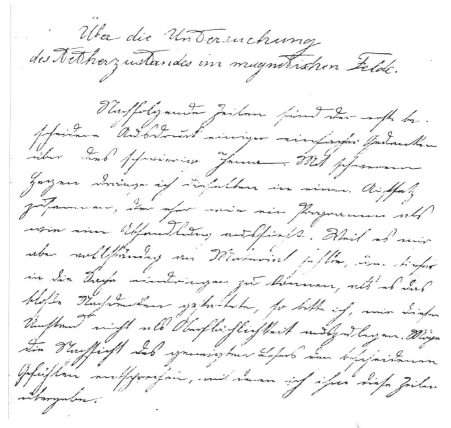

**Figure 1**. The first page of Einstein's early "scientific work" on the ether and magnetic field entitled *Über die Untersuchung des Ätherzustandes im magnetischen Felde*. The paper was written in 1894 or 1895 and sent to his uncle Caesar Koch. (Reproduced with the kind permission of The Albert Einstein Archives, The Jewish National and University Library, The Hebrew University of Jerusalem, Israel.)

his thoughts on the state of the ether in the magnetic field, and he sent his piece by mail to his uncle, Caesar Koch (his mother's brother), who lived in Belgium. The paper, written in 1894 or 1895, was published as a curiosity as late as 1971. It bears the title "A Study of the State of the Ether in the Magnetic Field."[22] When he wrote this text, Einstein was convinced of the existence of the ether, which he understood in his own specific way because he had no access to sources and materials that might have enabled him to carry out more thorough research and improve the presentation of the problem, instead of basing the work on his own

---

[22] A. Einstein, "Über die Untersuchung des Ätherzustandes im magnetischen Felde," *PhB*, 27, (1971) pp. 390-391; also in *Collected Papers of A. Einstein...* Vol. 1 (1987) pp. 6-9.

youthful speculations. This fact was mentioned in the introductory part of his paper.[23]

When he wrote his first work, Einstein did not yet identify the ether with real physical space, but he conceived of it as an elastic medium filling space and propagating all kinds of electromagnetic waves (light, heat, *etc.*). To him, ether meant a material medium with a certain mass, the density of which could be subject to minor fluctuations. As he assumed that elastic strains in elastic media do not produce considerable changes in their densities, he also claimed that one should expect only minor changes in the ether's density when dealing with its strains.

As he formulated his thoughts about the ether for the first time, Einstein was not yet convinced that the velocity of light should be constant in a free ether that was not influenced by other bodies. The idea of the constancy of the velocity of light was still quite alien to him. In his opinion, the velocity of waves propagating through the ether could change under the influence of three-component elastic forces that emerged when the ether was subjected to elastic strains.

As far as the magnetic field was concerned (actually the main topic of his paper), Einstein regarded it as representing potential states of the ether. His article contained a programme of experimental research on these states. He believed, however, that quantitative research (on the ether's density and elasticity, for example) could be started only after one had formulated certain qualitative ideas about its nature. In more general terms, at this point in his career Einstein felt that experiments must be supported by a good theory.

In order to learn about electromagnetic phenomena, the young "scholar" assumed that it was important to carry out appropriate experiments on the potential states of the ether in magnetic fields, *i.e.*, to perform adequate measurements of elastic strains in the ether as well as related forces of elastic deformation. Einstein believed that an elastic strain at any point in a given direction in the ether can be derived from the change in the velocity of the ether's wave at that point and in that direction. He claimed that the velocity of the wave should be directly proportional to the mass of the ether set in motion by these forces.

If changes in the length of an ether wave in a magnetic field could be experimentally detected, then we would be able to see whether the velocity of the wave was influenced by the component of the elastic

---

strain in the direction of the wave's propagation alone, or if this velocity was affected by transversal components of the strain as well.

Here is the main conclusion drawn by Einstein, with which he ends his first paper:

> Above all it must be demonstrated that there exists a passive resistance to the electric current due to the production of a magnetic field, which is proportional to the length of the current's path and is independent of the cross section and the material of the conductor.[24]

In discussing this conclusion, A. Pais points out that in his first paper young Einstein made an independent discovery of the qualitative properties of self-induction.[25]

These are the teenage Einstein's main conceptions of the ether. It is easy to see that, although they leave much to be desired in terms of their scientific value, these concepts suggest the ingenuity and inventiveness of the future scientist.

## 1.3.   Einstein designs experiments to confirm the Earth's motion through the ether

Einstein believed in the existence of the ether even during the first years of his studies at the Technical University of Zürich, Switzerland, where in the years 1896-1900—after finishing his classes at Swiss secondary school in Aarau—he attended courses in physics and mathematics intended for teachers.[26] At that time he had various ideas about possible experiments to confirm the Earth's movement with respect to the ether—experiments which he never carried out because of the sceptical attitude of his professors in general, and Professor H. F. Weber's negative position in particular.[27]

During the lecture delivered in 1922 at the University of Kyoto, Japan, explaining how he arrived at his Special Theory of Relativity, Einstein discussed one of his ideas:

> When I first thought about this problem, I had no doubt about the existence of the ether or the motion of the earth through it. I thought of the following experiment using two thermocouples: Set up mirrors so that the light from a single source is to be reflected in two different directions, one parallel to the motion of the earth and the other

---

[24] *Ibid.*, p. 391.

[25] A. Pais, *op. cit.*, p. 131.

[26] *Ibid.*, p. 41.

[27] *Ibid.*, pp. 45 and 132.

antiparallel. If we assume that there is an energy difference between the two reflected beams, we can measure the difference in the generated heat using two thermocouples. Although the idea of this experiment is very similar to that of Michelson, I did not put this experiment to a test.

While I was thinking of this problem in my student years, I came to know the strange result of Michelson's experiment. Soon I came to the conclusion that our idea about the motion of the earth with respect to the ether is incorrect, if we admit Michelson's null result as a fact. This was the first path which led me to the special theory of relativity. Since then I have come to believe that the motion of the earth cannot be detected by any optical experiment.[28] **b2**

Another experimental concept was mentioned by Einstein in a letter to his friend Marcel Grossmann, which he wrote in September 1901:

A new and considerably simpler method for the investigation of the motion of matter with respect to the luminiferous ether has come into my mind. It is based on the usual interference experiments. If only once inexorable destiny will allow me to finish with the necessary time and calm! When we meet again, I will tell you all about that.[29] **b3**

## 1.4.    Einstein's first doubts about the existence of the ether and the electrodynamics of his time

By the time he wrote the above words to Grossmann, Einstein had ceased to be fully convinced about the existence of the ether. He had doubts, but—as one might guess—he wanted to confirm or refute them by performing suitable experiments. Two years earlier, in a letter to Mileva Marić (his then fiancée who later became his first wife; she was also a physics student), he expressed his doubts with regard to the contemporary state of the electrodynamics of moving bodies and the existence of the ether:

I am more and more convinced that the electrodynamics of moving bodies, as it is presented today, does not correspond to reality, and that it will be possible to formulate it in a simpler way. The introduction of the term "ether" into the theories of electricity led to the notion of a medium, the motion of which one can discuss

---

[28] A. Einstein, "Speech at Kyoto University, December 14, 1922," *Physics Today*, 25 (1982), pp. 45-47.

[29] A. Einstein, "Letter to M. Grossmann, 6?/9/1901," *EA*, 11-485.

without, as I believe, being able to attribute a physical meaning to this statement.[30] **b4**

It is significant that this text included the title and touched the main problem of his paper of 1905 entitled "On the Electrodynamics of Moving Bodies" in which he presented his Special Theory of Relativity for the first time. We may conclude that the Special Theory of Relativity had been hatched and was maturing in Einstein's mind as early as 1899.

## 1.5    Conceptual premises for doubts concerning the ether

During his studies in Zürich (1896-1900), and also after his graduation, when he started his work at the Patent Office in Bern (1902), Einstein read a great deal and used to have discussions with his friends and colleagues. Influenced by his readings and discussions with Michelangelo Besso (lifelong friend from 1897) and others, Conrad Habicht and Maurice Solovine in particular (with whom he established a discussion group called the Olympus Academy), Einstein modified and developed his ideas with regard to the ether and space. The origins of these changes can be found in contemporary physics and philosophy, especially in positivism. What interests us here is that Einstein, as a student and later as an employee of the Patent Office, was particularly influenced by the works of Dutch physicist H. A. Lorentz, German physicist P. Drude, Austrian physicist and philosopher E. Mach, German chemist and philosopher W. Ostwald, philosopher R. Avenarius, and mathematician and philosopher Henri Poincaré.

In his student days, Einstein was greatly impressed by Lorentz's achievements in the field of electromagnetism. Einstein learned of these achievements by studying Drude's works and in 1895 by reading Lorentz's dissertation entitled *Toward a Theory of the Electrical and Optical Phenomena of Moving Bodies*.[31] The fact that Einstein studied this work in his student days was mentioned in the lecture at the University of Kyoto, quoted above.[32] Einstein greatly appreciated Lorentz's achievements and in his own works often referred to them as a giant step forward compared to Maxwell's achievements.

Lorentz's dissertation of 1895 exerted a significant influence on Drude, who, on adopting Lorentz's ideas, spread them in a distorted

---

[30] A. Einstein, "Letter to M. Marič, July 1899?" *EA*, FK-53.

[31] H. A. Lorentz, *Versuch einer Theorie der elektrischen und optischen Erscheinungen in bewegten Körpern* (Leiden: Brill, 1895).

[32] A. Einstein, "Speech at Kyoto University, December 14, 1922," *Physics Today*, 25 (1982), pp. 45-47.

way—or perhaps we should rather say that he simplified Lorentz's concept of the ether. Drude treated the ether as space having physical properties and remaining at absolute rest. That was not how the ether was originally understood by Lorentz. Although Lorentz presumed that the ether was at rest, there was no reason—as A. J. Kox has proved by his research[33]—to state that the ether was in a state of absolute rest. Lorentz only claimed that some parts of the ether remained at a standstill with respect to one another, and that the ether at rest constituted a privileged reference system. Moreover, he attributed a certain substantiality to the ether, which was not done by Drude. Here we quote an excerpt form Drude's *Handbook of Optics*, published in 1900. Einstein studied the book in depth and accepted Lorentz's ideas, apparently unaware that what he adopted was, in fact, only a distorted interpretation of Lorentz's concept of the ether:

> We will now hypothesise that the ether remains always at rest. On this basis H. A. Lorentz has developed a really complete and elegant theory, which is essentially the basis of the description presented here. The idea of an absolutely resting ether is the most simple and natural one, if one understands by ether not a substance, but only space endowed with certain physical characteristics.[34] **b5**

As we can see, the ether at rest constituted a privileged frame of reference for both Lorentz and Drude. In spite of his great respect for Lorentz's achievements, Einstein as a student could not accept a single distinguished frame of reference, as it seemed to introduce a specific kind of asymmetry. On the one hand, according to the laws of mechanics, all inertial reference frames are equivalent. On the other hand, Lorentz's electromagnetism was based on a privileged frame of reference. Einstein wanted to remove the asymmetry, as he believed that all inertial frames of reference should be equivalent from the viewpoint of electromagnetism. To a certain extent the result of Michelson's experiment, which Einstein learned from Lorentz's work (*cf.* §§ 89-92[35]), helped him to solve the problem. A solution without a special frame of reference seemed to be much simpler. He therefore supposed that a solution along these lines could be found, and said so in the above-quoted letter to Mileva Marić.

[33] A. J. Kox, "Hendrik Antoon Lorentz, the Ether and the General theory of Relativity," *AHESc*, vol. 38 (1988), nr 1, pp. 67-78.

[34] P. Drude, *Lehrbuch der Optik* (Leipzig: S. Hirzel, 1900), pp. 419-420.

[35] H. A. Lorentz, "Der Interferenzversuch Michelsons," in: H. A. Lorentz, *Versuch einer Theorie der elektrischen und optischen Erscheinungen in bewegten Körpern* (Leiden: Brill, 1895), para. 89-92.

Einstein described how he grappled with the problem in his lecture in Kyoto:

> I had a chance to read Lorentz's monograph of 1895. He discussed and solved completely the problem of electrodynamics to a first approximation, namely neglecting terms of order higher than $v/c$, where $v$ is the velocity of a moving body and $c$ is the light velocity. Then I tried to discuss the Fizeau experiment on the assumption that the Lorentz equations for electrons should hold in the frame of reference of the moving body as well as in the frame of reference of the vacuum as originally discussed by Lorentz. At that time I firmly believed that the electrodynamic equations of Maxwell-Lorentz were correct.
>
> Furthermore the assumption that these equations hold in the reference frame of the moving body leads to the concept of the invariance of the light velocity, which, however, contradicts the addition rule of velocities in the framework of mechanics.
>
> Why do these two concepts contradict each other? I realised that this difficulty was really hard to resolve. I spent almost a year in vain trying to modify the idea of Lorentz in the hope of resolving this problem.[36]

As can be seen, the main reason for Einstein's doubts about the existence of the ether originated in contemporary physics, or—to be exact—in contemporary electrodynamics, which distinguished between a special frame of reference associated with the ether, on the one hand, and all other frames, on the other hand. The asymmetry introduced by this distinction led Einstein to challenge the existence of the ether itself.

The works of Paul Drude, published at the end of the nineteenth century in Germany, exerted some influence on the young Einstein. We know for sure, for example, that Einstein read the above-mentioned *Handbook of Optics,* because upon reading it he wrote a letter to the author in which he offered his comments on the book.[37]

It is to be assumed that, as he was interested in the problems of the ether and electromagnetism, Einstein must also have read Drude's *Physics of the Ether Based on Electromagnetism,*[38] which appeared in 1894, and other works by Drude. These publications contained ideas that were developed

---

[36] A. Einstein, "Speech at Kyoto University, December 14, 1922," *Physics Today*, 25 (1982), pp. 45-47.

[37] According to Professor Don Howard, this letter has not yet been found.

[38] P. Drude, *Physik des Aethers auf Elektromagnetischer Grundlage* (Stuttgart: F. Enke, 1894).

by Einstein in his own way some time later on, after 1916, as the vocabulary and definitions introduced by Drude were then used in Einstein's articles on the relativistic ether. Similarities between expressions, and even identical ways they were used, offer proof that Einstein studied these works thoroughly. In his subsequent works Einstein would define the ether as "physical space endowed with physical attributes." Definitions like this can be found in the works by Drude, who identified the ether with real physical space, *e.g.*, when he wrote:

> Just as one can attribute to a specific medium, which fills space everywhere, the role of intermediary of the action of forces, one could do without it and attribute to space itself those physical characteristics which are now attributed to the ether. Until now people were afraid to adopt this conception, because the word "space" was associated with an abstract idea lacking physical properties.[39] **b7**

> The word "ether" does not imply any new hypothesis, but is only the essence of space free of matter, which has certain physical properties.[40] **b8**

Einstein, like Drude, would claim that there was no difference between physical space endowed with physical attributes and the ether. Although he used the same terminology as Drude, he did not understand the expression "physical space" in the sense of a space at absolute rest. Rather, Einstein regarded space as something ultra-referential, something to which no state of motion could be attributed, not even a state of rest.

Thanks to Drude's publications, Einstein was drawn into the realm of ideas that were eventually to bear fruit in the form of his modified theory of the ether. At that time, however, denying the existence of the ether suited his purpose much better than identifying the ether with real physical space. Under the influence of positivistic philosophy Einstein recognised physical space, which he identified with absolute space, as a metaphysical insertion.

Einstein was also heavily influenced by Mach's dissertation *The Science of Mechanics, a Critical and Historical Account of its Development*,[41] which was published in 1883 and read by Einstein in 1897. Einstein's attention was drawn to the book by his friend Besso:

In a letter dated April 8, 1952, Einstein wrote to Carl Seelig:

---

[39] *Ibid.*, p. 9.
[40] P. Drude, *Die Theorie in der Physik* (Leipzig: S. Hirzel, 1895), p. 9.
[41] E. Mach, *The Science of Mechanics, a Critical and Historical Account of its Development* (La Salle (Il.)/London: Open Court, 1942).

My friend Besso called my attention to Ernst Mach's *[The Science of Mechanics]*, when I was a student circa 1897. The book's critical attitude toward the fundamental conceptions and fundamental laws left a deep and lasting impression on me.[42] **b9**

Einstein learned about Mach's point of view, and about his theory of cognition in particular, not only from the above-mentioned *Mechanics...*, but also from the book *Analysis of Sense Impressions*. Einstein studied both books, and he discussed them with his friends at meetings of the Olympus Academy. This fact was mentioned by Solovine in his introduction to the published collection of his correspondence with Einstein.[43]

The cognition theory of Mach spoke of "pure experiments," "pure descriptions," and science "free of metaphysics." According to Mach, contemporary physics was not free of metaphysics, as it incorporated numerous interpolations that constituted supplementary constructions resulting from the operation of the mind alone. The human mind tends to treat as experience even those elements that are not reducible to experience at all. Notions such as "force," "matter," "atom," "absolute space" are our subjective inventions, not something experimentally tangible. They should therefore be eliminated from physics. After this, we would be left only with "sense impressions," which Mach preferred to call "elements." What we call the world is nothing but the system of such "elements."[44]

According to the plan for eliminating metaphysical interpolations, the inertia of bodies was to be considered with respect to the masses distributed over the whole Universe which could be attainable in the experiment, rather than with respect to absolute space, which was experimentally inaccessible. The inertia of bodies is something induced by distant masses, Mach asserted. This point of view was to be implemented by the young Einstein as a programme called "Mach's principle."

At this point it should be noted, however, that in spite of his epistemological views, Mach never rejected the existence of the ether; he admitted its hypothetical existence because he needed the ether as a

---

[42] A. Einstein, "Letter to C. Seelig, 8/4/1952," in: G. Holton, *Thematic Origins of Scientific Thought* (Cambridge, Mass.: Harvard University Press, 1973), p. 223.

[43] Maurice Solovine, "Introduction" in: A. Einstein, *Briefe an Maurice Solovine*, (1906-1955) (Berlin: Veb deutscher Verlag der Wissenschaften, 1960), p. 9.

[44] W. Tatarkiewicz, *Historia filozofii*, 5th edition, vol. 3 (Warsaw: Państwowe Wydawn. Naukowe, 1958), p. 132.

medium to transmit interactions between distant masses. He maintained that:

> [...] in the future we will learn more about this hypothetical medium, which scientifically is something more valuable than the doubtful notion of the absolute space.[45] **b9a**

From the fourth edition of his *Mechanics...* (1901) onward, Mach quoted the work of Paul Gerber who, in his research on the movement of Mercury's perihelion, proved that gravitational interactions propagated at the speed of light. He noted that

> ...this fact favours the hypothesis that the ether is the medium in which this propagation occurs.[46] **b10**

Mach's views on the ether and the reasons why he tended to accept its existence are presented in detail by J. Illy.[47]

The young Einstein, delighted with the cognition theory of Mach and, as we shall see, with other positivist views as well, nonetheless found reasons to doubt the existence of the ether, and then to reject it completely. The ether of Lorentz's electromagnetism, identified with absolute space at absolute rest—as Einstein regarded it under the influence of Drude's works—became something metaphysical in his eyes, *i.e.*, experimentally inaccessible, in conformity with the epistemological opinions of the positivists which influenced him so much at that time.

Many years later in his "Autobiographical Notes," Einstein wrote about his initial fascination with Mach's epistemological views:

> I see Mach's greatness in his incorruptible skepticism and independence; in my younger years, however, Mach's epistemological position also influenced me very greatly, a position which today appears to me to be essentially untenable. For he did not place the essentially constructive and speculative nature of thought, and more especially of scientific thought, in the correct light; in consequence of which he condemned theory on precisely those points where its constructive-speculative character invariably comes to light, as for example in the kinetic atomic theory.[48] **b11**

---

[45] E. Mach, *op. cit.*

[46] *Ibid.*

[47] J. Illy, "Mach, Lorentz and Einstein on Ether," in: *Ernst Mach and the Development of Physics*, Intern. Conf., Sept. 14-16, 1988, Prague.

[48] A. Einstein, "Autobiographisches," in: *Albert Einstein als Philososoph und Naturforscher*, ed. P.A. Schilpp (Braunschweig/Wiesbaden: Friedr. Vieweg und Sohn, 1979), p. 8.

Einstein abandoned Mach's epistemological views in the 1920s when he created his own theory of physical cognition, which anticipated the views of future philosophers of physics, such as Karl Popper.

The second edition of Ostwald's *Handbook of General Chemistry* was published in 1893.[49] We know with certainty that Einstein studied the book, as it inspired him to write his first research paper entitled "Consequences of the Phenomenon of Capillarity." Einstein mentioned the paper in his letter of March 19, 1901 to Ostwald,[50] in which he asked for a job in Ostwald's laboratory in Leipzig after completing his studies. He also made this evident in the paper, noting in the body of the text—and not in the footnotes—that much of his data were reproduced from Ostwald's *Handbook*.[51]

In the book, Ostwald presented his views on the ether, which testify to Ostwald's positivist epistemological attitude, since he, like Mach, was a positivist. Like Mach, Ostwald was also one of the leading opponents of the mechanistic interpretation of physical phenomena, and in his theory of cognition he linked sensorialism to energetism, as Mach did. Ostwald rejected the existence of the ether on the grounds that its attributes could not be directly observed, and he explained radiant energy—in accordance with his energetism—as a reality endowed with independent existence. Here we quote a few sentences from Ostwald's handbook:

> In the interest of a description of nature possibly free of hypothesis, we must ask if the acceptance of this medium, the ether, is unavoidable. It seems to me not to be the case.

> I cannot try to develop the previously given indications into a complete theory of light; it was only important for me to indicate the possibility of a pure energetic treatment of light. The main point is, after having recognised energy as a real substance—better, the only real substance of the so called external world—that we no longer have a need to search for a carrier of it, when we encounter it. This allows us to recognise radiant energy as existing autonomously in space.[52]
> **b12**

---

[49] W. Ostwald, *Lehrbuch der allgemeinen Chemie*, in zwei Bänden (Leipzig: W. Engelmann, 1893).

[50] A. Einstein, "Letter to W. Ostwald, 19/3/1901," in: H.G. Körber, "Zur Biographie des jungen Albert Einstein...," *Forschungen und Fortschritte*, 38, 1964, pp. 75-76.

[51] A. Einstein, "Folgerungen aus den Kapillaritätserscheinungen," *AdP*, 4 (1901), pp. 513-523.

[52] W. Ostwald, *op. cit.*, Bd. 2, 1Tl., pp. 1014 and 1016.

The idea of treating the energy of light as something self-sustained in space appeared in many of Einstein's papers after 1905, when he already had his own reasons, based on his Special Theory of Relativity (such as the superfluity of absolute space, equivalence of energy and mass $E = mc^2$), to deny the existence of the ether. We must admit, however, that even before all these ideas took shape, Ostwald's *Handbook of General Chemistry* had strengthened Einstein's doubts as to the existence of the ether, rather than weakened them. Gerald Holton, in particular, has noted the influence Ostwald's book had on Einstein's views.[53]

Avenarius, who held a professorship in Zürich from 1876, was, like Mach and Ostwald, a representative of the so-called "second positivism," which was different both from the "first positivism" of Auguste Comte, John Stuart Mill, Henry Thomas Buckle, and Herbert Spencer, and from the "third positivism" represented by the "Vienna Circle" (Moritz Schlick, Rudolf Carnap, Otto Neurath, and A. J. Ayer), the latter also known as logical positivism or neo-positivism.

One of the most important works by Avenarius, the *Critique of Pure Experience*,[54] was published at the turn of 1889-90. Einstein acquainted himself with the book in 1903 during the meetings of the "Olympus Academy." Reading and discussing a few chapters of the book with his friends must somehow have reinforced Einstein's positivist views, which he had earlier adopted from Mach and Ostwald.[55] Hence the "second positivism" must have had quite a strong impact on the young Einstein, a fact that came to light later on when he was working on his Special and General Theories of Relativity. Einstein distanced himself from positivism, started criticising it, and finally rejected it completely after formulating the General Theory of Relativity.

The representatives of all three branches of positivism, Mach especially, regarded David Hume as their predecessor. During the Olympus Academy meetings, Einstein studied Hume's *Treatise of Human Nature*[56] in detail. He also had thorough discussions with his friends on the *System of Logic* by John Stuart Mill, particularly the 3rd Chapter dealing

---

[53] G. Holton, "Mach, Einstein und die wissenschaftliche Suche nach Realität," in: G. Holton, *Thematische Analyse der Wissenschaft, Die Physik Einsteins und seiner Zeit* (Frankfurt am Main: Suhrkamp Verlag, 1981), pp. 203-254.

[54] R. Avenarius, *Kritik der reinen Erfahrung* (Leipzig: Fues [R. Reisland], 1888-1890).

[55] Maurice Solovine, *op. cit.*, p. X.

[56] *Ibid.*

with induction.[57] As a result, Einstein was exposed to the ideas of the representatives of the "first positivism."

The Olympus Academy held its afternoon meetings, which usually ran late into the night, virtually every day from Easter 1902 to November 1905, when Solovine moved to France to study at the University of Lyon. Most probably at the end of that period (as Solovine mentioned Poincaré's *Science and Hypothesis*[58] at the very end of the list of books discussed in the group[59]) Einstein discovered the world of ideas of Poincaré, who was a recognised predecessor of the theory of relativity, so that Whittaker,[60] together with Lorentz, considered Poincaré to be the creator of the Special Theory of Relativity.

Poincaré's *Science and Hypothesis*, which was a collection of excerpts taken from his various works written at the turn of the century, was published in 1902. Some of the works[61] represent the Special Theory of Relativity, which was just emerging, in a qualitative way, *i.e.*, without a mathematical formalism. The Olympus Academy spent considerable time discussing the book, and Solovine, a participant in the discussions, wrote:

> Poincaré's *Science and Hypothesis* is the book that highly impressed us and took our breath away for many weeks to come.[62] **b13**

In his book Poincaré also expressed his doubts about the existence of the ether. Unsuccessful attempts to detect the Earth's motion with respect to the ether (Michelson-Morley experiment) had undermined Poincaré's belief in the ether's existence. In his view, one could speak only about relative movements of certain bodies with respect to others, not about an absolute movement with respect to the hypothetical ether.[63] In Poincaré's book Einstein found support for his own doubts, which he had expressed as early as 1899 in a letter to his fiancée. It should be added that Mileva Marić participated in the meetings of the Olympus Academy from the day of her marriage with Einstein (January 6, 1903). Years later Solovine wrote: "Mileva, intelligent and reserved, listened to us attentively, but never took the floor in a discussion."[64]

---

[57] *Ibid.*

[58] H. Poincaré, *La science et l'hypothèse* (Paris: Flammarion, 1968).

[59] Maurice Solovine, *op. cit.*, p. XIV.

[60] E. Whittaker, *op. cit.*, vol. 2, pp. 27-77.

[61] H. Poincaré, *op. cit.*, chapters X and XIV.

[62] Maurice Solovine, *op. cit.*, p. X.

[63] H. Poincaré, *op. cit.*, pp. 180-182.

[64] Maurice Solovine, *op. cit.*, p. XIV.

To sum up this chapter, we once again stress that Einstein was inclined to doubt the existence of the ether, and finally rejected it for physical reasons (such as the above-mentioned asymmetry and the results of the Michelson-Morley experiment) as well as philosophical reasons (due to a significant impact from the "second positivism"). These two types of reasons become apparent when we analyse works by Einstein in which he denies the existence of the ether.

# Chapter 2

# EINSTEIN DENIES THE EXISTENCE OF
# THE ETHER (1905-1916)

Einstein denied the existence of the ether for 11 years only, *i.e.*, from 1905 to 1916. This chapter presents all his works from that period in which the issue of the ether was mentioned and discussed in any way. It also presents the history of the formulation of the General Theory of Relativity starting in 1907: a history that brought about a gradual discovery of real attributes of space which were not perceived by Einstein until 1916. In this process, subconscious for Einstein, a new concept of the ether was born.

## 2.1.  Works published before the Special Relativity Theory

The year 1905 brought a vast number of papers by Einstein: he published five articles in *Annalen der Physik* and thirteen reviews in *Beiblätter zu den Annalen der Physik*.[65] This was the year the notion of the light quantum was introduced, the Special Theory of Relativity was finally formulated, and Einstein was awarded a doctoral degree for his dissertation *A New Method for Determining Dimensions of Molecules*.[66] It was also the year when the word "ether" appeared in Einstein's academic work for the first time in the article in which he introduced the notion of

---

[65] Reviews by A. Einstein in: *Beiblätter zu den Annalen der Physik*, 29 (1905), pp. 235-238, 240-242, 246-247, 623-624, 629, 634-636, 640-641, 950, 952-953, 1114-1115, 1152-1153, 1158.

[66] A. Einstein, *Eine neue Bestimmung der Moleküldimensionen* (Bern: Wyss, 1905), also in: *AdP*, 19 (1906), pp. 289-306.

a light energy quantum. This was a well-known and very important paper,[67] and eventually the ideas and results developed brought him the Nobel Prize. The ether was mentioned here incidentally, quite superficially, without any discussion of what it was, or if it existed at all. Using contemporary terminology, Einstein mentions incidentally the existing division into matter and the ether that was commonly used at that time. He does so in the first section of his article, where he points to certain difficulties encountered by the theory then in use of perfect blackbody radiation. On this basis, Einstein found a dependence expressed by a mathematical equation which was a condition of dynamic equilibrium of the radiation. He stated, however, that:

> This relation, found as a condition for dynamical equilibrium, is not only in disagreement with experience, but it also means that in our description we cannot speak of a well-defined distribution of energy between ether and matter.[68] **c1**

This is how Einstein characterised the difficulty resulting from the use of the contemporary theory of perfect blackbody radiation, which did not recognise the existence of light energy quanta. By introducing the quanta he not only eliminated the problem, but also explained the photoelectric effect. At the same time Philipp Lenard was conducting experiments dealing with the same effect, for which he was later awarded the Nobel Prize.

Einstein used the results of his research when he denied the existence of the ether as an argument to treat radiant energy (in the same way Ostwald had) as an independent entity which did not require a carrier in the form of the ether, because self-existent quanta of energy do not need any carrier at all.

The German physicist Lenard defended a specific kind of ether to the end of his days. This is why the two Nobel laureates, Einstein and Lenard, who truly respected each other initially (as proved by their correspondence[69]), entered into a strictly scientific controversy. In the period when Nazism began to emerge their hostility became open, and Lenard attacked Einstein's theory of relativity on political and anti-Semitic grounds, which had nothing to do with science proper. Lenard became the chief founder of the so-called German physics, defending the

---

[67] A. Einstein, "Über einen die Erzeugung und Verwandlung des Lichtes betreffenden heuristischen Gesichtspunkt," *AdP*, 17 (1905), pp. 132-148.

[68] *Ibid.*, p. 136.

[69] A. Kleinert, Ch. Schönbeck, "Ph. Lenard und A. Einstein, Ihr Briefwechsel und ihr Verhältnis vor der Nauheimer Diskussion," *Gesnerus*, 35 (1918), pp. 318-333.

existence of the ether, which opposed the so-called Jewish physics that included, first of all, Einstein's theory of relativity with its rejection of the ether as such. But let us go back to 1905, the year when the Special Theory of Relativity was finally formulated.

## 2.2    The ether becomes superfluous

Einstein took a just five weeks to complete the formulation of the final version of the Special Theory of Relativity. In Kyoto, Einstein mentioned the fact that he was helped by his friend Besso. We quote a further excerpt from his lecture delivered in 1922. We have already shown that Einstein had been trying for a year to alter Lorentz's theory in such a way that he could remove the contradiction between the classical rule of speed addition and constant speed of light. Einstein found it necessary to remove this contradiction, because this would help him to remove the asymmetry discussed in the previous chapter.

> By accident a friend of mine in Bern (M. Besso) helped me out. It was a beautiful day when I visited him with this problem. I started the conversation with him in the following way: "Recently I have been working on a difficult problem. Today I come here to battle against that problem with you." We discussed every aspect of the problem. Then suddenly I understood where the key to the problem lay. Next day I came back to him again and told him without even saying hello: "Thank you. I've completely solved the problem." An analysis of the concept of time was my solution. Time cannot be absolutely defined and there is an inseparable relation between time and signal velocity. With this new concept, I could solve all the difficulties completely for the first time.
>
> Within five weeks the special theory of relativity was completed.[70]

In his "On the Electrodynamics of Moving Bodies,"[71] in which Einstein published his Special Theory of Relativity, there are two comments on the ether, and the word "ether" itself is used only once. Both comments can be found in the introduction to the article. The first is as follows:

> Examples of this sort, together with the unsuccessful attempts to discover any motion of the earth relatively to the "light medium," suggest that the phenomena of electrodynamics as well as of

---

[70] A. Einstein, "Speech at Kyoto University, December 14, 1922," *Physics Today*, 25 (1982), pp. 45-47.

[71] A. Einstein, "Zur Elektrodynamik bewegter Körper," *AdP*, 17 (1905), pp. 891-921.

mechanics possess no properties corresponding to the idea of absolute rest.[72] **c3**

Here Einstein mentions the results of the Michelson-Morley experiments. He did not, however, mention the names of the two physicists—just as Poincaré had failed to mention them in his report presented at the international conference of physicists in Paris in 1901. The report was published in the collection of Poincaré's works *Science and Hypothesis*, which Einstein knew. We quote an excerpt from the report:

> [...] a great deal of research has been carried out concerning the influence of the Earth's movement [with respect to the ether—L.K.]. The results were always negative.[73] **c4**

Did Einstein consciously imitate Poincaré? This eventuality must be excluded in this case. We can, however, assume that by not mentioning the names of the two experimenters, Einstein avoided having to quote their article, and Einstein hated writing footnotes. He considered this time-consuming task a waste of time, and did not like to quote even his own works. Moreover, as he had already noted in his 1905 paper, the results of Michelson's experiments were—to some extent—of secondary importance. The primary emphasis was on the above-mentioned asymmetry which he removed, and Einstein based his theory upon two assumptions:

1) equivalence of all inertial systems for formulating laws of both mechanics and electrodynamics and optics;
2) constancy of light speed in these systems independent of the movement of the source of light.

The very first sentence of Einstein's article hints at the existence of the asymmetry in contemporary theories of electromagnetism:

> It is known that Maxwell's electrodynamics—as usually understood at the present time—when applied to moving bodies, leads to asymmetries which do not appear to be inherent in the phenomena.[74] **c5**

It should be pointed out once again that Einstein's first comment about the ether quoted above shows clear influence of the positivistic

---

[72] *Ibid.*, p. 891. English translation in: A. Einstein, H. A. Lorentz, H. Weyl, H. Minkowski, *The Principle of Relativity* (New York: Dover, 1952). The quoted sentence is on p. 37.

[73] H. Poincaré, *La science et l'hypothèse* (Paris: Flammarion, 1968), p.182.

[74] A. Einstein, in: *Zur Elektrodynamik...*, p. 891.

cognitive theory of Mach, Ostwald, and Avenarius. Einstein indicated that "no attribute characteristic of the phenomena corresponds to the notion of absolute rest;" in other words, absolute rest remains beyond any experimental test, "not only in mechanics, but also in electrodynamics." Since it lacked the typical attribute of a phenomenon, it became a metaphysical interpolation which had to be eliminated from physics.

The other comment by Einstein about the ether is as follows:

> The introduction of a "luminiferous ether" will prove to be superfluous inasmuch as the view here to be developed will not require an "absolutely stationary space" provided with special properties, nor assign a velocity-vector to a point of the empty space in which electromagnetic processes take place.[75] **c6**

As we can see, Einstein found the luminiferous ether unnecessary, because his theory, based upon the assumptions set out above, did not need a space at an absolute rest characterised by any specific attributes. The existence of such a space would contradict his first assumption, which he would call the Special Principle of Relativity.

It must be stressed that in both his comments on the ether, Einstein had in mind nothing other than Lorentz's ether in a form distorted or simplified by Drude. This ether was identified with "space at absolute rest characterised by specific attributes," and a velocity vector had to be attributed to points of this ether when electromagnetic processes were considered in the reference system moving with respect to the absolute space.

Having pointed out to the reader in the introduction to this article that the ether proved unnecessary under the presented arguments, Einstein stopped discussing it in the later part of his work. With this silence he treated the problem of the ether as absolutely settled.

In his paper "Does the inertia of a body depend on its energy content?" (1905),[76] Einstein demonstrated the dependence between the mass of a body and the energy contained in it ($m = E/c^2$) for the first time as a result of his freshly formulated Special Theory of Relativity. The article did not include a single mention of the ether; it is mentioned here, however, because Einstein used these conclusions in his further works (in which he denied the existence of the ether) as an argument for treating

---

[75] *Ibid.*, p. 38.

[76] A. Einstein, "Ist die Trägheit eines Körpers von seinem Energieinhalt abhängig?" *AdP*, 18 (1905), pp. 639-641.

the energy of radiation with its relativistic mass $(m = E/c^2)$ as an independent entity which did not need a carrier in the form of the ether. This argument had its foundations in the conclusion he reached and expressed in the last line of the work:

> If the theory corresponds to the facts, the radiation conveys inertia between the emitting and the absorbing bodies.[77] **c7**

Having treated the ether as something absolutely superfluous, Einstein made no further mention of it in any of his works of the years 1905–1907. At the end of 1907, however, Einstein had to break his silence about the ether, because it was not easy to convince everyone that the ether was unnecessary. For example, Einstein's works failed to convince Lorentz that the ether was superfluous; Lorentz defended the concept of the ether until the end of his life, and his tenacity led Einstein to introduce a new concept of the ether in 1916.

Einstein broke his silence about the ether only in order to reject it once again. On December 4, 1907 the scientific magazine *Jahrbuch der Radioaktivität* received Einstein's long article titled "The Principle of Relativity and its Consequences."[78] The word "ether" appeared four times in the article: three times in the introduction, and once in the first part, devoted to relativistic kinematics. From both the context and the stress Einstein twice laid on the word we can see that he had in mind an ether absolutely at rest, as represented by Lorentz's theory. The ether appeared along with an explanation of the principle of relativity, which constitutes the first of the two basic assumptions of his Special Theory of Relativity.

This work was used by Einstein—as he indicated in the introduction—to unify all the results of his Special Theory of Relativity contributed not only by himself, but also by Max Planck, Max von Laue, and Kurd von Mosengeil. These include achievements in the field of kinematics, electrodynamics, mechanics of a material point, and mechanics and thermodynamics of physical systems. These achievements, as Einstein showed, resulted from an appropriate integration of Lorentz's theory with the principle of relativity, according to which there was no privileged system within the class of inertial reference systems used to describe the laws of mechanics, optics, and electrodynamics. The integration was possible thanks to a precise analysis

---

[77] *Ibid.*, p. 71

[78] A. Einstein, "Relativitäts prinzip und die aus demselben gezogenen Folgerungen," *JR*, 4 (1907), pp. 411-462.

of the concept of time. It turned out that the so-called "local time" introduced by Lorentz, which was used as auxiliary quantity in order to reconcile the theory with the results of the Michelson-Morley experiment, had to be treated as time in a given system of reference. Each system has its own time, and absolute time does not exist.

Then it turned out that fundamental equations of Lorentz's theory correspond to the relativity principle of the Special Theory of Relativity, and that the Lorentz-Fitzgerald hypothesis on bodies contracting in the direction of their motion, introduced *ad hoc* in order to reconcile the theory with experiment, proved to be a natural consequence of the new theory which simultaneously established the relative (not absolute) nature of the contraction. In the theory of relativity which was created only by reconciling Lorentz's theory with the relativity principle, only an ether perfectly at rest proved to be absolutely unnecessary. We quote Einstein's words on this accomplishment:

> Only the idea of a luminiferous ether as a carrier of the electric and magnetic forces does not fit the present theory; because electromagnetic fields do not appear here as states of something material, but as independently existing things that are similar to ponderable matter and that have in common with it the characteristic of inertia.[79] **c8**

As we can see, the physical and philosophical arguments formulated by Einstein in his paper of 1905 are joined by a new physical argument based upon the principle of equivalence of mass and energy which was one of the chief consequences of the new theory. Electromagnetic fields did not need a carrier because they were something independent, since they have a property of inertia similar to ponderable matter.

## 2.6    Beginnings of the General Theory of Relativity

The 1907 article represented a summary of all the results of the Special Theory of Relativity reached so far, as well as a conclusion of a certain stage of experiments and the beginning of a new stage. The Special Theory of Relativity formulated by Einstein covered mechanical, optical, and electromagnetic phenomena, but it never touched upon the problem of gravitation. It dismissed the privileged reference system identified by Lorentz with the ether at absolute rest, but still favoured only one class of reference systems, *i.e.*, the class of inertial systems. It became clear to Einstein that the next stage of research should also

---

[79] *Ibid.*, p. 413.

remove this distinction. The relativity principle should also apply to all kinds of non-inertial systems.

After summing up the results of the Special Theory of Relativity in four parts of his article, in the fifth and final part, called "Relativity Principle *versus* Gravitation,"[80] Einstein tried for the first time to extend the relativity principle to uniformly accelerated non-inertial systems. It was the first time Einstein had tried to draw attention to the fact that a non-inertial reference system in uniformly accelerated motion, *i.e.*, with constant acceleration $\gamma$ is physically equivalent to an inertial system with a homogeneous gravitational field characterised by gravitational acceleration $\gamma$. In both systems all bodies, irrespective of their shape and composition, will fall in vacuum with the same acceleration $\gamma$. In two such systems, gravitational phenomena appear identical, as do mechanical, optical, and electromagnetic phenomena. These phenomena are equally influenced by the field of acceleration in one system and the field of gravitation in the other system.

It was the first time that Einstein had shown how the gravitational field and acceleration influenced the spatial shape of bodies, the working of clocks, and electromagnetic processes. As regards electromagnetic processes, the influence of both fields on light propagation was presented for the first time. The velocity of light changes according to the relations

$$c = c_o\left(1 + \frac{\Phi}{c_o^2}\right); \quad c = c_o\left(1 + \frac{\gamma h}{c_o^2}\right)$$

(where $\Phi$ is Newton's scalar gravitational potential, $\gamma h$ is an equivalent of the potential in the field of acceleration, and $c_o$ is the constant velocity of light in inertial systems without a gravitational field), and the light ray is subject to curving by the formula $\left(\gamma / c_o^2\right)\sin\varphi$ (where $\varphi$ is an angle between the direction of gravitational force and the light ray). As regards the impact of the gravitational potential on the working of the clock, we may expect a shift toward red of the spectrum of radiation from massive bodies. The atoms emitting radiation must thus be treated as a specific kind of clock.

It should be noted that the influence of the gravitational field and acceleration upon the phenomena, outlined above, as described in Einstein's paper, meant the beginning of a new concept of the ether, which Einstein failed to notice as yet. Physical space turned out to have a

---

[80] *Ibid.*, pp. 454-462.

structure, which truly influenced physical phenomena, thus becoming something real, with real physical attributes that could influence the behaviour of bodies and the course of processes occurring within it.

Einstein was still with the Patent Office in Bern when he came to the idea of equal treatment of uniformly accelerated systems of reference and systems within which there was a homogeneous gravitational field.[81] He was to call this idea the luckiest discovery of his life.[82]

## 2.7     Einstein finds a new argument against the ether

On June 17, 1907 Einstein applied for a position as a *Privatdocent* at the University of Bern. His application (with his seventeen publications and CV enclosed) was rejected, however, as he failed to fulfil the *Habilitationsschrift* or postdoctoral lecturing qualification required for the position. As a result, Einstein had to write a new dissertation to satisfy these requirements. On February 18, 1908 he was awarded his *venia legendi* and took up the position of *Privatdocent* at the University. However, he continued to work at the Patent Office, as his university post was unpaid.[83] The *Habilitationsschrift* has never been published and is lost. We only know its title: "Consequence Resulting from the Distribution of the Blackbody Radiation for the Composition of Radiation."[84] Pais[85] believes that the results of this research were published in 1909 in Einstein's two works printed in *Physikalische Zeitschrift*: one in March[86] the other in October.[87] In the latter article, which he presented at the conference of physicists in Salzburg, titled "On the Development of Our Ideas on the Nature and Composition of Radiation," Einstein returned to the issue of the ether after two years of silence. This happened in the first part of the October article, where he presented the main ideas of his Special Theory of Relativity as an introduction to his improved theory on the dual—*i.e.* quantum and wave—nature of light. As we know, Einstein introduced the notion of the quantum of light as early as 1905. In the Salzburg paper he formulated his theory of the dual nature of light in clear terms. Light,

---

[81] A. Einstein, "Speech at Kyoto University, December 14, 1922," *Physics Today*, 25 (1982), pp. 45-47.

[82] A. Einstein, "Grundgedanken..." (*Morgan Manuscript*), *EA* 2070.

[83] A. Pais, *op. cit.*, p. 184-185.

[84] *Ibid.*, p. 185.

[85] *Ibid.*

[86] Einstein, "Zum gegenwärtigen Stande des Strahlungsproblems," *PhZ*, 10 (1909), pp. 185-193.

[87] A. Einstein, "Entwicklung unserer Anschauungen über das Wesen und die Konstitution der Strahlung," *PhZ*, 10 (1909), pp. 817-825.

in his opinion, had a double structure: a quantum structure (*Quantenstruktur*) and a wave structure (*Undulationsstruktur*). Although he admitted that the new theory lacked adequate mathematical support, in his opinion, light constituted singularities (*Singularitäten*) or singular points (*singularen Punkten*), which contained the total energy of the electromagnetic field, and were surrounded by the force field (*Kraftfeld*), which was a plane wave (*ebene Welle*).

Einstein used the Special Theory of Relativity, briefly presented in this paper, to prove that one of its consequences is the equivalence of energy and mass. He used this statement to show that light energy quanta carry mass from the emitting body to the absorbing body. For the first time Einstein was talking not only about Lorentz's ether, but also about the ether which was believed to move together with matter, or to be partly entrained by it. Therefore, Einstein briefly presented Armand Fizeau's experiment, which was designed to determine whether the ether was carried along with matter. When discussing the ether at rest, he briefly mentioned the experiment of Michelson-Morley, the result of which—according to Einstein—led to the admission of the relativity principle in electromagnetism. Thus, Einstein wrote:

> [...] more generally, and above all in relation to every system in motion without acceleration, [we wish to] proceed exactly according to the same laws. In what follows we will call this presupposition simply the "principle of relativity." Before we touch upon the question whether it is possible to remain faithful to the principle of relativity, we wish to reflect briefly about what will happen to the ether hypothesis if we maintain this principle.

> On the basis of the ether hypothesis, experiment leads us to regard the ether as at rest. The principle of relativity implies that every law of nature referred to a coordinate system $K'$ moving uniformly relative to the ether is equivalent to the corresponding law referred to a coordinate system $K$ at rest relative to the ether. If this is so, we may equally imagine the ether at rest relative to $K'$, as relative to $K$. It is then completely unnatural to distinguish one of the coordinate systems $K$ and $K'$ and introduce an ether at rest relative to it. From this it follows that a satisfactory theory can only be reached if we renounce the ether hypothesis. The electromagnetic fields that constitute light no longer appear as states of a hypothetical medium, but as autonomous forms that are emitted from light sources, just as in Newton's theory of light emission. As in the latter theory, a space

not crossed by radiation and without ponderable matter appears to be really empty.[88] **c9**

In the above excerpt we note two familiar arguments which Einstein used to deny the existence of the ether. The first argument was that when we distinguished one reference system by introducing a notion of the ether into it, we had an unnatural asymmetry that violated the principle of relativity. Electromagnetic fields constituted autonomous formations which did not need a carrier in the form of the ether—that was the other argument. Both arguments were of a physical nature. A new physical argument was the use of the corpuscular and wave theory of light (which had just recently been formulated in Einstein's dissertation) to deny the existence of the ether. In Einstein's opinion, the energy of light and its related inertia (relativistic mass) were carried by autonomous quanta, which—as in the Newtonian emission theory of light—did not need a carrier.

Is it possible that Einstein knew the Newtonian corpuscular-wave theory of light? Did he mean this theory when he spoke of emission theory? It is true that Newton formulated the first corpuscular-wave theory of light; he is known, however, as the creator of the purely corpuscular theory, contrasted with later wave theories. The Newtonian corpuscular-wave theory had been long-forgotten, and it was almost completely unknown in Einstein's time. The theory is included in the second volume of the English edition of *Optics*.... The book contains, among other things, a very interesting method for easily inducing reflection and penetration of light; in order to explain the phenomena of reflection and penetration of light when thin plates were used, Newton found it necessary to introduce an element of periodicity into the theory of propagation of light. He imagined that corpuscles of light excite a certain oscillation period in the surrounding the ether waves. These waves precede corpuscles and adjust them for easy reflection or penetration.[89] According to Newton, heat radiation could also reach areas from which the air had been pumped out (the experiment with two thermometers, one of which was placed in vacuum, the other in the air) not only because the region was penetrated by radiation particles, but also because of the existence of the ether, which is set in undulating motion by these particles.[90]

---

[88] *Ibid.*, p. 819.

[89] L. Rosenfeld, "Newton's Views on Aether and Gravitation," *AHESc*, 6 (1969), pp. 29-37, see pp. 34-35.

[90] H. Guerlac, "Newton's Optical Aether," *NRRSL*, 22 (1967), pp. 45-57.

It is probable that Einstein did not know Newton's corpuscular-wave theory at all, and when he was talking about Newtonian theory of emission, he meant the purely corpuscular theory attributed to Newton. Einstein noticed that according to the theory, space without light and matter was absolutely void. The Newtonian corpuscular-wave theory had assumed the existence of the ether, which was contradicted by the theory of Einstein, and it would therefore be absolutely pointless to refer to a theory that assumed the existence of the ether to support a theory that denied its existence.

When Einstein presented his new theory of light at the conference of physicists in Salzburg, he was neither an employee of the Patent Office nor a *Privatdocent* at the University of Zürich. He gave up both positions when he was offered the position of assistant professor of theoretical physics at the University of Zürich. He assumed his duties on October 15, 1909. On October 22 he moved into his new flat at 12 Moussonstrasse in Zürich together with his wife Mileva, and his son Hans Albert.[91]

## 2.8    Other works in which Einstein rejects the ether

In the years 1905-1916, it was in his "Relativity Principle and Its Consequences in Contemporary Physics"[92] that Einstein dealt exhaustively with the old ether, and finally rejected it. This paper was published in 1910 in French in the magazine *Archives des sciences physiques et naturelles*. It was a compilation of abstracts from two papers already discussed here (*cf.* footnotes 14 and 22 to this chapter), linked into a new whole and complemented by elements referring mainly to the old ether.

The problem of the old ether and the need to reject it were discussed in the first five points of this article, such that the first part of the article was devoted entirely to the old ether and arguments for its non-existence. Thus, none of the new arguments against it were to appear here.

The further works[93] written by 1916 that mentioned the ether included merely repeated historical data and the reflections and arguments Einstein had used in the article presently under our scrutiny, and thus we can treat the issue as a whole, discussing all aspects together.

---

[91] A. Pais, *op. cit.*, pp. 185-186.

[92] A. Einstein, "Principe de relativité et ses consequences dans la physique moderne," *ASPN*, 29 (1910), pp. 5-28 and 125-244.

[93] A. Einstein, "Relativitätstheorie," *VNGZ*, 56 (1911), pp. 1-14; by the same author "Relativitätstheorie," in: *Die Physik*, ed. E. Lecher (Leipzig: Teubner, 1915), pp. 702–713.

In his further works Einstein presented a more or less developed historical sketch of the best-known hypotheses and models of the ether, and a brief description of the experiments of Fizeau and Michelson-Morley. Then he expressed his enthusiasm for the results of Lorentz's electromagnetism, but criticised Lorentz, noting that his ether at absolute rest contradicted the relativity principle. In Einstein's opinion, Lorentz had to accept the existence of an ether at absolute rest owing to his specific interpretation of experiments; at the same time, none of the experimental facts undermined the relativity principle as applied to classical mechanics. Thus, Einstein could see a solution to the problem (mentioned above, as he had written about it in his earlier papers) of reconciling Lorentz's electromagnetism with the relativity principle. This reconciliation was linked to a rejection of the hypothetical ether.

> The first step to take if one wants to attempt this reconciliation is to renounce the ether.[94] **c10**

Both basic assumptions of the Special Theory of Relativity, *i.e.*, the relativity principle, and the principle of constancy of light speed, forced one to abandon the hypothesis of an ether in a state of absolute rest. Einstein stressed this in his article "The Theory of Relativity,"[95] published in the collective work *Physics*:

> It is easy to understand why we had to renounce introducing a light-carrying ether in the theory. In fact, if every ray of light is to propagate in the vacuum with velocity *c* relative to *K*, we must imagine this light-carrying ether everywhere at rest with respect to *K*. But if the laws of propagation of light relative to the system *K'* (moving with respect to *K*) are the same relative to *K*, we must with equal right accept the existence of a light-carrying ether at rest with respect to *K'*. Since it is absurd to accept that the ether is at rest at the same time with respect to both systems and since it would be only slightly less absurd to prefer one of the two (better, one of an infinite number of) physically equivalent reference systems in the theory, we must renounce the introduction of this idea, which in any event was a useless ornament of the theory, since we had already abandoned a mechanical interpretation of light.[96] **c11**

Einstein had always seen clearly that the hypothesis of an ether at rest could never agree with both postulates of his Special Theory of

---

[94] A. Einstein, "Principe de la relativité...," p. 19.

[95] A. Einstein, "Relativitätstheorie," in: *Die Physik*, ed. E. Lecher (Leipzig: Teubner, 1915), pp. 702–713, see p. 708.

[96] *Ibid.*, p. 708.

Relativity, therefore, when he delivered his lecture "The Theory of Relativity" on January 11, 1911 at the meeting of the Zürich Natural Society, he once again pointed out:

> The theory outlined in the following is not compatible with the hypothesis of an ether.[97] **c12**

## 2.9 Origin of the dispute between Einstein and Lenard

As was mentioned in Section 2.1, Einstein and Lenard at first respected and admired each other for their results in research on the photoelectric effect. Lenard was conducting experiments on photoelectricity, which were to earn him the Nobel Prize in 1905, and Einstein provided a theoretical explanation by introducing the notion of the light energy quantum, for which he received the Nobel Prize in 1921. The two scholars exchanged letters about their research. Einstein maintained a correspondence with Jakob Laub, who was Lenard's assistant from 1908. Laub visited Einstein in Bern, where he stayed for some time. Einstein and Laub were bound by ties of friendship, which bore fruit in the form of three joint papers. On the basis of the existing correspondence between Einstein and Lenard (and also Laub) we are able to reconstruct a history of their contacts. This was actually done by A. Kleinert and Ch. Schönbeck in their *Lenard and Einstein, Their Correspondence and Interrelations before the Discussion in Nauheim in 1920*.[98] We limit ourselves to a brief presentation of their results in the years 1905-1916.

At first, Lenard who had held the position of full professor at the University of Kiel since 1891, did not know Einstein. The latter knew of Lenard at least from reading the papers he referred to clearly in his article on light quanta published in 1905.

> As far as I can see our point of view is not in contradiction with the properties of the photoelectric effect observed by Mr. Lenard.[99] **c13**

Einstein might have sent Lenard his article, although we do not know with any certainty if he did. Lenard, however, must have read it, because soon after it had been published he sent Einstein his own latest work.[100] Einstein expressed his gratitude for the gesture in his letter of November 16, 1905, where he also included his comments on the subject

---

[97] A. Einstein, "Relativitätstheorie," *VNGZ*, 56 (1911), p. 2.

[98] A. Kleinert, Ch. Schönbeck, *op. cit.*

[99] A. Einstein, "Über einen die Erzeugung...," p. 147.

[100] Ph. Lenard, "Über die Lichtemission der Alkaldämpfe und Salze und über die Zentren dieser Emission," *AdP*, 17 (1905).

matter of Lenard's paper. Lenard did not reply immediately, as he was of a different opinion. However, he kept thinking about Einstein's comments. Four years passed before Einstein received an answer to his letter.

In the meantime, Laub had become Lenard's assistant. In the correspondence exchanged between Einstein and Laub we can see that Einstein greatly appreciated Lenard's experimental results. In his letter of congratulations on Laub's appointment as Lenard's assistant, Einstein described Lenard "as a great master and an original mind."

> Dear Mr. Laub, First of all my warmest congratulations for your position of assistant and for your salary. I took much pleasure in this news. But I think that the opportunity to collaborate with Lenard is much more than a position of assistant and a salary put together! Try to tolerate his eccentricities, however many he has. He is a great master and an original mind! Perhaps he will be sociable to a man he has learned to respect.[101] **c14**

and two years later, on March 16, 1910 he also wrote to Laub:

> You must be happy because of your staying with Lenard, particularly because—as it seems to be—you are able to treat him with great dexterity. He is not only a skilful master in his field, but also a true genius."[102] **c15**

Laub informed Einstein that Lenard also appreciated him for his work on the photoelectric effect. With the knowledge and agreement of Lenard, Laub was also concerned with the Special Theory of Relativity, and he published an article titled "Experimental Foundations of the Relativity Principle."[103]

On June 5, 1909, Lenard replied to Einstein's letter of November 16, 1905 from his new place of residence and his new job at the University of Heidelberg. In his letter, he called Einstein a great and wide-ranging thinker, and he stated that he kept Einstein's letter on his desk while he was in Kiel, and still kept it on his desk in Heidelberg. Einstein's letter had stimulated his thinking, which took time to ripen, and that was the reason for the delay in his reply, for which he apologised, and invited Einstein to visit him at any convenient moment.

---

[101] A. Einstein, "Letter to J. J. Laub (undated) 1908," in: A. Kleinert, Ch. Schönbeck, *op. cit.*, p. 320.

[102] A. Einstein, "Letter to J. J. Laub 16/3/1910," in: A. Kleinert, Ch. Schönbeck, *op. cit.*, p. 320.

[103] J. J. Laub, "Über die experimentalen Grundlagen des Relativitätsprinzip," *JR*, 7 (1910), pp. 405-420.

Einstein's attitude toward Lenard changed radically after Lenard delivered his paper *Ether and Matter*[104] at the Academy of Science in Heidelberg on June 4, 1910. Lenard defended the concept of the ether. In his opinion, the ether was necessary to explain numerous phenomena, particularly electromagnetism and gravitation. In Lenard's view, the ether was not a continuous substance, but something discrete. It consisted of parts (*Ätherteile*) which he called cells (*Zellen*). Their rotational movement was the reason why material objects could move through it without visible resistance. Lenard failed, however, to give any mathematical description of his model of the ether. In his lecture he declined to mention Einstein's name, nor did he mention his Special Theory of Relativity. He mentioned the relativity principle, however, stressing that it had already appeared in Galileo's works. He did agree that on the basis of mechanical phenomena, it was impossible to discover absolute movement, and only motion of bodies relative to others was perceivable. However, he opted for absolute motion, which was confirmed by electromagnetic phenomena. He claimed that Lorentz had explained the negative results of Michelson's experiment as due to contraction of bodies in the direction of their motion when they move against the ether. Lenard's expanded paper was published in the form of a book in 1910 and 1911.

Einstein reacted violently against it in his letter to Laub. In his violent criticism he used words hardly translatable into English. We quote them first in their original version: "*Lenard muss aber in vielen Dingen sehr 'schief gewickelt' sein. Sein Vortrag von neulich über die abstruse Ätherei erscheint mir fast infantil.*"[105] **c16** (Lenard must be very misguided in many things. His recent contribution on the abstruse ideas of ether seems to me almost infantile.)

Three years later, in 1913, when Lenard was planning to set up a chair of theoretical physics in Heidelberg, unaware of Einstein's critical attitude towards him, he wrote to Sommerfeld, that he would eagerly create such a professorship "if a personality like Einstein [...] could be available."[106]

---

[104] Ph. Lenard, *Über Äther und Materie*, Zweite, ausführlichere und mit Zusätzen versehene Auflage (Heidelberg: C. Winters Universitätsbuchhandlung, 1911).

[105] A. Einstein, "Letter to J. J. Laub, undated," in: A. Kleinert, Ch. Schönbeck, *op. cit.*, p. 322.

[106] Ph. Lenard, "Letter to A. Sommerfeld, 4/9/1913," in: A. Kleinert, Ch. Schönbeck, *op. cit.*, p. 322.

From that moment until 1917, there is no trace of any contact between the two scholars.[107] The further history of the controversy between Einstein and Lenard will be discussed later in this study.

## 2.10  Minkowski's four-dimensional world

During his Zürich studies, Einstein was for a time a disciple of an eminent mathematician, Hermann Minkowski, who is renowned for giving Einstein's Special Theory of Relativity a new mathematical form. The idea of fusing time and space into a single, four-dimensional system of events complying with Lorentz's transformations emerged in Minkowski's mind as early as 1905, although it was not published until two years later. On November 5, 1907, Minkowski delivered a paper at the Mathematical Society in Göttingen (where he moved from Zürich in the same year). It was published later in *Annalen der Physik* under the title "The Principle of Relativity."[108] A month later Minkowski presented his work titled "Basic Equations for Electromagnetic Processes in Moving Bodies"[109] The paper contained a detailed mathematical development of the four-dimensional world. Most famous, however, was his lecture "Space and Time," delivered on September 21, 1908 in Cologne at the 80th Gathering of German Naturalists and Physicians, which was published in 1909 in *Physikalische Zeitschrift*. Minkowski began his lecture with the words:

> The views of space and time which I wish to lay before you have sprung from the soil of experimental physics, and therein lies their strength. They are radical. Henceforth space by itself, and time by itself, are doomed to fade away into mere shadows, and only a kind of union of the two will preserve an independent reality.[110] **c17**

Max Born, who became Minkowski's disciple somewhat later than did Einstein, testified that the idea of the four-dimensional world took shape in Minkowski's mind soon after the Special Theory of Relativity was published. On July 16, 1955 at the congress on The 50th Anniversary

---

[107] A. Kleinert, Ch. Schönbeck, *op. cit.*, p. 322.

[108] H. Minkowski, "Das Relativitätsprinzip," *AdP* 47 (1915), pp. 927-938.

[109] H. Minkowski, "Die Grundgleichungen für die elektromagnetischen Vorgänge in bewegter Körper," *Gött. Nachr.*, 53 (1908). The reader can find more details in: T. Hirosige, "Theory of Relativity and the Ether," *Japanese Studies in the History of Science*, 7 (1968), pp. 37-53.

[110] H. Minkowski, "Raum und Zeit," *PhZ*, 10 (1909), p. 104-111. English transl. in: H. A. Lorentz, A. Einstein, H. Minkowski and H. Weyl, *The Principle of Relativity* (New York: Dover, 1952), pp. 75-91.

of Relativity Theory which took place in Bern, Max Born described a seminar on the theory of the electron held by Minkowski in the following words:

> My memory of these long bygone days is of course blurred, but I am sure that in this seminar we discussed what was known at this period about the electrodynamics and optics of moving systems. We studied papers by Hertz, Fitzgerald, Larmor, Lorentz, Poincaré, and others, but also got an inkling of Minkowski's own ideas which were published only two years later. [...] Minkowski published his paper "Die Grundlagen für die elektromagnetischen Vorgänge in bewegten Körpern" in 1907. It contained the systematic presentation of his formal unification of space and time into a four-dimensional "world" with a pseudo-Euclidean geometry, for which a vector and tensor calculus is developed.[111] **c18**

At first Einstein was not enthusiastic about the new mathematical description of his Special Theory of Relativity; in particular he found the use of tensors to be "superfluous erudition" (*überflüssige Gelehrsamkeit*); this was exactly the expression he used in his conversation with Valentine Bargmann,[112] his future co-worker. From 1912 Einstein started to use the tensor method himself, and from 1916 he recognised that he owed Minkowski a great deal, and that Minkowski's results and method greatly facilitated his transition from the Special Theory of Relativity to the General Theory of Relativity. Einstein mentioned the four-dimensional world of Minkowski (remarking that it was a positive development) in his works published for the first time in 1910, four years after the idea was first published.[113]

We might also add that in his works on the relativistic ether, Einstein stressed its four-dimensional nature, which was a consequence of Minkowski's concept of a four-dimensional expression of relativity theory.

## 2.11   Einstein on the path to the new ether

It was the General Theory of Relativity that contributed the greatest impetus to the introduction of the new ether. This new concept of the ether was Einstein's physical interpretation of the space-time continuum

---

[111] M. Born, "Physik und Relativität," in: M. Born, *Physik im Wandel meiner Zeit*, 4th edit. (Braunschweig: F. Vieweg u. Sohn, 1966), pp. 186 and 192. English transl. in: M. Born, *Physics in my Generation* (New York: Springer, 1969), pp. 101 and 106.

[112] A. Pais, *op. cit.*, p. 152.

[113] A. Einstein, "Principe de la relativité...," p. 139.

of the General Theory of Relativity. Therefore, we present below the history of the formulation of the theory and point out those elements which led to the introduction of the new concept of the ether.

In March 1911, Einstein moved, together with his family, from Zürich to Prague to accept the post of full professor of theoretical physics at the German University named after Karl Ferdinand. He worked at the university for eleven months only.[114] Of the articles written there, four were further attempts to generalise relativity theory and formulate a new theory of gravitation. The first was titled "On the Influence of Gravitational Force on Propagation of Light."[115] This article did not contain any important new thoughts compared to what was included in Section 5 of Einstein's 1907 paper. Certain new comments and qualifications in it, however, should be noted. Using a thought experiment, Einstein proved clearly and precisely that an increase in energy $\Delta E$ was accompanied not only by an increase in inertial mass of $\Delta E / c^2$, but also by an increase in gravitational mass (weight). The variability of light velocity in an acceleration field and in a gravitational field was presented in a more clear way: the velocity was a function of position, as it depended on the potential $\Phi$ (or $\gamma h$) at a given location. Einstein thus noted for the first time:

> The principle of constancy of the velocity of light is, according to this theory, not valid in the formulation which is usually taken as a basis for the ordinary theory of relativity.[116] **c19**

He also stressed the relativity of accelerated motion for the first time:

> According to this point of view one can as little speak of *absolute acceleration* of the reference frame as one can discuss of *absolute velocity* of a system within the usual theory of relativity.[117] **c20**

Because light velocity in a gravitational field was a function of position, the light ray was also bent in the field. In his article of 1907, Einstein noted that in terrestrial conditions the effect is so small that it cannot be experimentally perceived. He indicated that the effect would be much bigger in the neighbourhood of the Sun, and so there was a

---

[114] A. Pais, *op. cit.*, p. 193.

[115] A. Einstein, "Einfluss der Schwerkraft auf die Ausbreitung des Lichtes," *AdP*, 35 (1911), pp. 898-908.

[116] A. Einstein, "Über den Einfluss der Schwerkraft auf die Ausbreitung des Lichtes," *AdP*, 35 (1911), pp. 898-908.

[117] *Ibid.*, p. 906.

possibility of discovering it by astronomical measurements. The article ended with an appeal to astronomers to conduct appropriate experiments.

In another Prague paper, titled "Velocity of Light and the Static Field of Gravitation,"[118] Einstein for the first time put forward the hypothesis that an acceleration field could be treated as a specific case of gravitational field:

> [...] at least in my opinion, the hypothesis that the "acceleration field" is a special type of gravitational field has such a high probability [...][119]
> **c21**

The chief innovation of this paper was the following idea: Einstein proposed that the locally variable value of the light velocity should be treated as the quantity (scalar) which characterised the gravitational field at a given position. In the case of a static gravitational field, we could write a linear equation of the field (which would correspond to Poisson's equation used in the theory of gravitation then in use) in the following way:

$$\Delta c = kc\rho$$

where $k$ was a constant related to the gravitational constant $G$ as follows:

$$k = \frac{G}{c^2},$$

and $\rho$ was the density of matter.

In this work Einstein continued to utilise flat space, although—for the first time—he was aware that in the new theory of gravitation he would have to depart from this notion of flatness. This flatness, for instance, did not occur

> ... in a uniformly rotating system, in which due to the Lorentz contraction the ratio of the circumference to the diameter should be different from $\pi$, as a consequence of our definition of length.[120] **c21'**

Einstein utilised flat space in his next Prague work as well, and that paper, titled "On the Theory of the Static Field of Gravitation,"[121] was a

---

[118] A. Einstein, "Lichtgeschwindigkeit und Statik des Gravitationsfeldes," *AdP*, 38 (1912), p. 355-369.

[119] *Ibid.*, p. 355.

[120] *Ibid.*, p. 356.

[121] A. Einstein, "Zur Theorie des statischen Gravitationsfeldes," *AdP*, 38 (1912), pp. 443-458.

continuation of the previous one. In it, Einstein attempted to show the influence of the gravitational field on electromagnetic and thermal phenomena, and he introduced differential equations for the field, as he noticed a certain weakness in the equations in his previous paper. Newton's principle of action and reaction was the source of the problems, as it turned out that the consistent application of the equation $\Delta c = kc\rho$, equivalent to Poisson's equation, together with the equation

$$F = -\text{grad } c$$

expressing the force $F$ influencing a unit volume of matter density $\rho$, violated the principle of action and reaction.

The title of Einstein's fourth Prague paper is a question: "Does There Exist a Gravitational Interaction that is Analogous to Electromagnetic Induction?"[122]

One of the results of this paper, which Einstein found quite interesting, was a theoretical confirmation of the increase of the mass of one body in the proximity of another. This result led, according to Einstein, to a more general conclusion, that the mass of any material point is definitely the product of the interaction of all the other masses of the universe. This conclusion was equivalent, Einstein noted, to the postulate used by Mach in his *Mechanics*.

Pais believes[123] that the fact that Einstein published his fourth article in a magazine, the *Vierteljahresschrift für gerichtliche Medizin*, which was not well known among physicists, meant that he was not sure about his assumptions in reference to the induced force of gravitation. It must be stressed, however, that the idea that particle masses are the product of an interaction with all the other masses in the universe would figure in Einstein's research work for many years.

His Prague works constituted a step forward—although Einstein completely failed to notice it—towards the new ether. Space in the Special Theory of Relativity without matter and electromagnetic field appeared, as Einstein remarked, to be absolutely empty, or in other words, without physically perceptible features. Presently, with the introduction of the equivalence principle, a system in uniformly accelerated motion appeared to have the features of the field of gravitation. It could really influence the working of a clock, affect measuring rods, change the velocity of light, and bend light rays. Thus, it

---

[122] A. Einstein, "Gibt es eine Gravitationswirkung die der elektrodynamischen Induktionswirkung analog ist?," *VgM*, 44 (1912), pp. 37-40.

[123] A. Pais, *op. cit.*, p. 204.

was no longer possible to say that space was physically neutral, that it was a void lacking any physical features. Einstein was not yet aware of this, however, since attributing physical characteristics to space was philosophically alien to him; he was still under the strong influence of the second positivism, in which space was a metaphysical interpolation that should be removed from physics. Einstein wanted to implement Mach's programme in which the emergence of the acceleration fields with the motion of certain systems, and the emergence of centrifugal forces with rotary motion, should not be attributed to the influence of the specific nature of the structure of space, which was physically imperceptible, and therefore did not constitute an entity in the physical sense of the word. Rather, these fields and forces were due to interactions with the fixed stars, which constitute real physical entities.

The nineteenth century brought the discovery of non-Euclidean geometries. Janos Bolyai and Nikolai Lobachevski were the creators of negative curvature geometry, which describes the surface of a funnel or a horse saddle in two dimensions. Carl Friedrich Gauss and Bernhard Riemann created positive curvature geometry, which we use in two-dimensional form on the surface of a sphere or ellipsoid. Riemann also created a geometry covering all three types: plane geometry of zero-curvature, and the positive and negative curvature geometries. The surface of Klapp's favourite fruit, the pear, has all three curvatures: it has regions of positive curvature, which are fragments of a sphere, negative curvature in the funnel-like regions, and very small flat zones. Two-dimensional Riemannian geometry allows a description of this surface, since it is two-dimensional. Riemann's metric geometry is so general that it can be used to describe space with any number of dimensions. Within this geometry, the most important role is played by the square form

$$ds^2 = \sum g_{\mu\nu} dx_\mu dx_\nu$$

which defines the metric of space and constitutes a generalisation of the Pythagorean theorem, a generalisation which also refers to spaces of non-zero curvature. The symbol $ds$ signifies the distance between indefinitely close spatial points (so-called linear element), $dx_\mu$ and $dx_\nu$ are differentials of co-ordinates introduced into the space of the co-ordinate system, and $g_{\mu\nu}$ is a basic metric tensor, the components of which are functions of co-ordinates. When we can introduce the system of co-ordinates in which components of the tensor $g_{\mu\nu}$ differ from unity and zero, we are dealing with the space of positive or negative curvature. The $g_{\mu\nu}$ tensor describes the metric behaviour of space at a given location.

In Einstein's works on the new ether we generally do not encounter any mathematical formulae, as was mentioned in the Introduction. Two formulae, a symbol of a basic metric tensor, and the symbols generalising Pythagorean theorem of square form, do appear. We shall learn later why this is the case: for now, we will simply answer the question: Why did Einstein start to use Riemann's geometry?

As we know, in his second Prague paper, Einstein remarked that the proportion of the perimeter to the diameter of a platform in rotation due to Lorentz contraction was a value different from $\pi$, which meant a departure from Euclidean geometry. This fact must have been one of the reasons why Einstein started to look for mathematical tools which could be used to formulate a new theory of gravitation in non-Euclidean geometries. During his stay in Prague, Einstein often held discussions with fellow mathematician George Pick about his attempts to formulate a new theory of gravitation. Pick drew Einstein's attention to the fact that he could find a mathematical model to formulate his ideas in the works of Ricci and Levi-Civita on Riemann's geometry.[124] Pick's suggestions met with no response, as Pais notes,[125] probably because the absolute differential calculus developed by the two Italian mathematicians seemed too abstract to Einstein. Einstein tended to be more concrete in his way of thinking.

At the end of his stay in Prague, Einstein remembered the lectures in Gauss's geometry he had attended as a student at the Technical University in Zürich, which were given by Professor Carl Friedrich Geiser. The knowledge he gleaned there pertained to more concrete matters: Gauss's geometry referred to surfaces of positive curvature. Gauss's co-ordinate system was seen by Einstein as a key to solving any difficulties he encountered. He learned from his friend Grossmann that the geometry of Gauss had been developed into a multi-dimensional case by Riemann. Here is what Einstein said on this subject in Kyoto in 1922:

> Describing physical laws without reference to geometry is like describing our thoughts without words. We need words in order to express ourselves. What should we look for to describe our problem? This problem was left unsolved until 1912, when I hit upon an idea that the surface theory of Gauss might be the key to solve this

[124] Ph. Frank, *Einstein. Sein Leben und seine Zeit* (Wiesbaden: Friedr. Vieweg u. Sohn, 1979), p. 141.

[125] A. Pais, *op. cit.*, p. 212.

mystery. I found that Gauss's surface coordinates were very helpful in understanding this problem. Until then I did not know that Riemann had discussed the foundations of geometry more deeply. I happened to remember the lecture on geometry in my student years by Professor Geisser who discussed the Gauss theory. I found that the foundations of geometry had deep physical meaning in this problem.

When I came back to Zürich from Prague, my mathematician friend Grossmann was waiting for me. He had helped me before in supplying mathematical literature when I was working at the patent office in Bern and had some difficulties in obtaining mathematical articles. First he taught me the work of Ricci and later the work of Riemann. I discussed with him whether the problem could be solved using the Riemann theory, in other words using the concept of the invariance of the line elements. We wrote a paper on this subject in 1913, although we could not obtain the correct equation for gravity.[126]
**c24**

And in 1923, he came back to this theme:

Einstein's second statement on the July-August period was made in 1923: "I had the decisive idea of the analogy between the mathematical problem of the theory [of general relativity] and the Gaussian theory of surfaces only in 1912, however, after my return to Zürich, without being aware at that time of the work of Riemann, Ricci, and Levi-Civita. This [work] was first brought to my attention by my friend Grossmann, *when I posed to him the problem of looking for generally covariant tensors whose components depend only on derivatives of the coefficients* [$g_{\mu\nu}$] *of the quadratic fundamental invariant* [$g_{\mu\nu}dx_\mu dx_\nu$]" [emphasis by Pais].[127]

Einstein's words (particularly the emphasised text) prove that he felt the basic square invariant $ds^2 = g_{\mu\nu}dx_\mu dx_\nu$ (linear element) and its coefficients $g_{\mu\nu}$ were the most basic element in formulating his new theory of gravitation.

The mathematician Marcel Grossmann, Einstein's friend from the days of their studies in Zürich at the Technical University, had been a tenured professor at the University since 1907, and in 1911 he was elected Dean of the Department of Physics and Mathematics. One of the first steps taken by the new Dean was to give Einstein a professorship in Zürich. Einstein and his family moved there from Prague in August

---

[126] A. Einstein, "Speech at Kyoto University, December 14, 1922," *Physics Today*, 25 (1982), pp. 45-47.

[127] A. Pais, *op. cit.*, p. 212.

1912. In Pais's opinion, during the period between August 10 and 16 that year it became obvious to Einstein that Riemann's geometry was the most suitable tool for the new theory of gravitation.[128] It should be noted, however, that pseudo-Riemann geometry served this purpose, as pseudo-Euclidean geometry did for the world of Minkowski. A four-dimensional time-space continuum differs from an ordinary four-dimensional geometry in the introduction of specific time coefficients. In both pseudo-geometries it is necessary to multiply one or three co-ordinates, depending on convention, by an imaginary number $i = \sqrt{-1}$, which means that the basic invariant d$s$ may also assume negative values—a phenomenon not encountered in the ordinary geometries of Euclid and Riemann.

The co-operation between Einstein and Grossmann resulted in two jointly written articles. The first was called "Outline of the Generalised Theory of Relativity and Theory of Gravitation."[129] The results of this publication were also presented in the form of two papers delivered by both authors at the general meeting of the Swiss Naturalist Society which took place in Frauenfeld on September 9, 1913. Einstein presented the physical aspects of the new theory, and Grossmann its mathematical apparatus. Einstein's paper was published under the title "Physical Foundations of a Certain Theory of Gravitation,"[130] and its *précis* was published under the title of "Theory of Gravitation."[131]

A brief introductory characterisation of the first of these articles is given by Michal Heller, who writes:

> This article is exactly what it promises in its title—the *outline* of the new theory, which also constituted a generalisation of the theory of relativity of 1905. The formulating of this draft was undoubtedly a moment of breakthrough. It became clear then that all the efforts made so far were a search—often intuitive and blind—for new concepts, and the attempt to combine them into the fragments of some larger unity. And all of a sudden everything "clicked" into the right place. It was obvious what the whole would look like, although it was not always clear which tools should be utilised to construct the whole. The remainder of the great adventure would be a path marked

---

[128] *Ibid.*, p. 210.

[129] A. Einstein and M. Grossmann, "Entwurf einer verallgemeinerten Relativitätstheorie und einer Theorie der Gravitation," *ZMPh*, 62 (1913), pp. 225-261.

[130] A. Einstein, "Physikalische Grundlagen einer Gravitationstheorie," *VNGZ*, 58 (1913), pp. 284-290.

[131] A. Einstein, "Gravitationstheorie," *VSNG* (1913), 2. Teil, pp. 137-138.

by dramatic mistakes and setbacks—yet everyone knew where it would lead.[132] **c24**

The joint work by Einstein and Grossmann contained two clearly separate parts: Grossmann's contribution was a clear presentation of Riemann's geometry and its tensor calculus. Grossmann started his part with a discussion of the invariability of the generalised linear element

$$ds^2 = \sum g_{\mu\nu}dx_\mu dx_\nu$$

when curvilinear systems of co-ordinates were introduced. Then he defined tensors and the basic operations of tensor algebra. In addition, his part included certain mathematical details intended to support Einstein's arguments.

The following is one of the main concepts in the part written by Einstein, which incorporated arguments from the paper delivered in Frauenfeld in which some of the ideas were presented—or perhaps emphasised—more clearly. In the theory of relativity known so far (the Special Theory of Relativity), which was based on the assumption of constant velocity of light in inertial reference systems, an isolated material point was in rectilinear motion with constant velocity according to the equation

$$\delta\left(\int ds\right) = 0$$

where

$$ds^2 = -dx^2 - dy^2 - dz^2 + c^2 dt^2$$

This is the equation of the motion of the material point in inertial systems of reference applied to the Special Theory of Relativity. When one moves from one inertial system into another, the linear transformations of Lorentz are applied. The space-time interval $ds$ is the invariant of these transformations.

Introducing the principle of equivalence creates a dependence of light velocity on the gravitational potential. The velocity of light proves variable in accelerated systems and in the field of gravitation. The linear transformations of Lorentz are no longer applicable to these reference systems. Because the relativity principle is generalised over all systems, the theory has to be formulated in such a way that the space-time interval

---

[132] M. Heller, "Jak Einstein stworzył ogólną teorię względności? (How did Einstein create the theory of general relativity?)," *PF*, 39 (1988), pp. 3-21; see pp. 8-9.

d$s$ becomes invariant under any arbitrary transformation. It is possible to construct such a theory by applying the absolute differential calculus used by Riemann, and developed by C. G. Ricci and Tullio Levi-Civita.

In the generalised theory of relativity in which the speed of light is a function of position, material points in free motion execute movements described by the equation

$$\delta\left(\int ds\right) = 0$$

where

$$ds^2 = \sum g_{\mu\nu} dx_\mu dx_\nu$$

This is the equation of the material point moving within any gravitational field. The generalised linear element d$s$ which appears in this equation is invariant under arbitrary transformations. The coefficients $g_{\mu\nu}$ constitute components of the covariant symmetric metric tensor. These components characterise the gravitational field in a given position, and— as Einstein noted in his paper delivered in Frauenfeld[133]—replace the scalar gravitational potential of the Newtonian theory of gravitation.

It must be pointed out that Einstein was still far from identifying the gravitational field (whose real structure is described in the new theory by the tensor $g_{\mu\nu}$) with real physical space possessing real metric structure. Einstein recognised space as something completely devoid of physical features because in his mind the notion of space was linked to systems of coordinates which became absolutely arbitrary and devoid of physical meaning in his General Theory of Relativity, now nearly mature. The epistemological ideas of Mach and the other positivists also had some influence on this conception. This seems assured by a letter Einstein sent to Mach, where he presents the results of his paper written in 1913 together with Grossmann. Here is an excerpt from the letter:

> For me it is absurd to attribute physical properties to "space." The totality of masses generates a field $g_{\mu\nu}$ (gravitational field) which controls the development of every process, including the propagation of light rays and the behaviour of measuring rods and watches. An event will first of all be referred to four *completely arbitrary* variables of space and time.[134] **c25**

---

[133] A. Einstein, "Physikalische Grundlagen ...," p. 286.
[134] A. Einstein, "Letter to E. Mach, undated," in: V. P. Vizgin, Ya. A. Smorodinskii, "From the equivalence principle to the equation of gravitation," *Sov. Phys. Usp.*, 22 (7), July 1979, pp. 489-515; see p.. 499.

This absolute arbitrariness of space-time variables made Einstein sure that the theory formulated by him deprived space of the "last remnant of reality." He made it clear in his further works, particularly in his paper, which contained definitive formulation of the General Theory of Relativity.

At that time—we must stress once again—Einstein insisted on linking the concept of space with the concept of reference systems and coordinate systems introduced into them, which became, in his new theory, absolutely arbitrary and devoid of physical meaning. Einstein treated this as a full implementation of Mach's epistemological programme, according to which—as we know—space, as a metaphysical interpolation devoid of physical features, ought to be removed from physics.

Later, when Einstein linked the notion of space with the carrier of metric structure, he acknowledged that due to this structure, space acquired measurable physical features, and he attributed real existence to it, calling it—due to those new features—the new ether. It was then that he started to identify the field of gravitation, understood as a physical object, with four-dimensional space also understood as a physical object, calling it "the field-$g_{\mu\nu}$, *i.e.* time-space continuum" (*cf.* **d54** in Appendix).

In their joint work, Einstein and Grossmann did not arrive at generally covariant equations of the gravitational field, although they were quite close to their goal. Due to a series of errors, two years were to pass before the final covariant equations of the gravitational field were found. A large number of publications have been devoted to this period of trial and error, and the final success. These include books and essays written by V. P. Vizgin,[135] A. Pais,[136] J. Stachel[137] and J. Norton.[138] Michal Heller also sketches a brief history of the period in his article.[139] We quote an excerpt in which he pinpoints where Einstein and Grossmann went wrong:

> Today, it is a textbook example (which still poses serious difficulties
> for beginners in the theory of relativity). Einstein and Grossmann had

---

[135] *Ibid.*, pp. 489-515.

[136] A. Pais. *op. cit.*, chapters 12, 13 and 14.

[137] J. Stachel, "Einstein's Search for General Covariance, 1912-1915," in: *Einstein and the History of General Relativity*, eds. D. Howard, J. Stachel (Boston-Basel-Berlin: Birkhäuser, 1989), pp. 63-100.

[138] J. Norton, "How Einstein found his field equations?," in: *Einstein and the History of General Relativity*, pp. 101-159.

[139] M. Heller, "Jak Einstein...," pp. 11-15.

no idea about Bianchi's identity and about the fact that, out of the ten components of the field equations proposed by them, only six could possess physical content. The remaining four only reflected freedom of selection of coordinate systems, and could be selected in practically any arbitrary way. No wonder Einstein and Grossmann "proved" that the components of the metric tensor (*i.e.*, the gravitational potentials) could not "be determined" by the equations, because they really cannot: four components can be chosen arbitrarily. And here we have a certain paradox: the "indetermination" of the equations results from their invariance (irrespective of the coordinate system selection). Einstein was seeking invariant equations and he found them, but he rejected them just because of one property which resulted from invariance (he was naturally unaware of it). The subject matter of Einstein's further research—because the relative success of his and Grossmann's publication satisfied him only for a short time—was in practical terms an attempt to understand the error made in 1913.[140]
c26

The Relativity Theory, and particularly the developing General Theory of Relativity, aroused considerable scepticism, if not acute objections from some physicists. In the magazine *Scientia* there appeared two articles criticising the theory of relativity. Einstein reacted with a rebuttal "On the Relativistic Problem."[141] In this essay he showed the extent to which he was inspired by Mach's cognitive theory and what he called Mach's principle. The following is a long excerpt from the rebuttal in which Einstein expressed, among other things, his disbelief in the real existence of space:

> Two masses very distant from all celestial bodies float in cosmic space. They are close enough to be able to exert forces on one another. An observer follows the movements of the two bodies, always looking in the direction of the line joining the two masses to the firmament of the fixed stars. He will observe that his line of sight cuts a closed line on the visible firmament of the fixed stars, which does not change its position, always with respect to the visible firmament of the fixed stars. If the observer has a natural intelligence but has studied neither geometry nor mechanics, he will conclude thusly: "My masses perform a motion which is at least in part causally determined by the system of the fixed stars. The laws according to which masses move in my surrounding are codetermined by the fixed stars." A man, who has completed scientific studies will smile at the naiveté of our observer and say to him: "The movement of your

---

[140] *Ibid.*, p. 10.
[141] A. Einstein, "Zum Relativitätsproblem," *Scientia*, 15 (1914), pp. 337-348.

masses has nothing to do with the sky of fixed stars; rather, it is determined by the laws of mechanics totally independently of the other masses. There is a space *R* in which these laws are valid. These laws are such that your masses remain constantly on the same plane of this space. The system of the fixed stars cannot rotate in this space, because it would be rent asunder by enormous centrifugal forces. It therefore remains necessarily at rest (at least approximately!) if it is to exist permanently; from this it follows that the plane on which your masses move always passes through the same fixed stars." But our fearless observer will say: "You are certainly learned beyond comparison. *But just as I was never convinced to believe in ghosts, I do not believe in this big thing, of which you are talking to me and that you call space. I can neither see anything like this, nor can I imagine anything of the kind..* [L.K. emphasis] Or should I imagine your space *R* as a very subtle but concrete net, to which the remaining objects are related? In this case besides *R* I can imagine a second net *R'* of the same type which is moving in an arbitrary way relative to *R* (for example it rotates). Are your equations equally valid relative to *R*?" The learned man denies this with certainty. After this the naive gentleman says: "How do the masses know relative to which of the "spaces" *R*, *R'*, *etc.* they must move according to your laws, from what do they recognise the space, or the spaces, in reference to which they must behave?" Now our learned man is in a great difficulty. He understands that there must exist privileged spaces of this type, but he cannot indicate any reason why these spaces should be distinguished from others. Then the naive gentleman says: "*In this case I provisionally consider your preferred spaces an idle invention and keep my opinion that the vault of fixed stars codetermines the mechanical behaviour of my experimental masses.*" [L.K. emphasis][142] **c27**

In his reply, Einstein used the results of the article written together with Grossmann, and he quoted only this publication. New events in Einstein's life brought an end to this stage of his co-operation with Grossmann, as Einstein was offered membership in the Prussian Academy of Sciences, and moved to Berlin on April 6, 1914. His wife and two sons left him soon after that, moving back to Zürich, and it was a shock to him. After she left him, he rented a bachelor flat at 13 Wittelsbacherstrasse.[143]

The Berlin years started with a major article titled "The Formal Foundation of the General Theory of Relativity."[144] In this publication

---

[142] *Ibid.*, pp. 344-345.

[143] A. Pais, *op. cit.*, pp. 224 and 240.

[144] A. Einstein, "Formale Grundlage der allgemeine Relativitätstheorie," *SPAW* (1914), 2. Teil, pp. 1030-1085.

Einstein summed up all the results leading to the new theory of gravitation. One novelty in the article was that for the first time Einstein introduced the geodesic equation as the equation motion of the point particle.[145] A large part of the article was devoted to a presentation of the tensor calculus he had learned through his co-operation with Grossmann.

Soon afterwards, their second joint article appeared: "Covariant Attributes of Field Equations in the Theory of Gravitation Constructed upon the Generalised Theory of Relativity."[146] This work did not contain the final formulation of the General Theory of Relativity either. It was simply an attempt to expand the class of transformations under which the equations of the gravitational field could be covariant.[147]

The local press was interested in Einstein's move to Berlin; among other requests, the Berlin newspaper *Die Vossische Zeitung* asked him for a short article explaining the theory of relativity to the lay public. As a result, on the twentieth day after Einstein arrived in Berlin, April 26, 1914, the paper published his article "On the Relativity Principle," in which Einstein characterised the rejection of the hypothesis of the ether as a major achievement of the relativity theory:

> Two of the main results of the theory of relativity will be mentioned here, which should also be interesting for the layman. The first is that the hypothesis of the existence of a space-filling medium to support the propagation of light (the luminiferous ether), must be abandoned. According to this theory light appears no longer to be a state of motion of an unknown carrier, but a physical object to which must be attributed a completely autonomous physical existence. The second is that the theory shows that the inertia of a body is not an absolutely invariable or constant, but it grows with the content of energy. The important conservation laws of mass and energy thus blend into a unique law; the energy of a body defines the mass itself at the same time.[148] **c28**

Einstein derived the generally covariant equations of gravitational field by himself, although a certain mutual inspiration from his contacts with renowned mathematician David Hilbert cannot be excluded. This

---

[145] M. Heller, "Jak Einstein...," p. 12.

[146] A. Einstein und M. Grossmann, "Kovarianzeigenschaften der Feldgleichungen der auf die verallgemeinerte Relativitätstheorie gegrundeten Gravitationstheorie," *ZMPh*, 63 (1914), pp. 215-225.

[147] M. Heller, "Jak Einstein...," p. 12.

[148] A. Einstein, "Vom Relativitätsprinzip," *Die Vossische Zeitung*, 26 April 1914, p. 1.

possible influence is pointed out by Jagdish Mehra.[149] Einstein was
staying in Göttingen at the end of June and the beginning of July 1914,
where, in six two-hour-long lectures, he presented his search for the new
theory of gravitation. In the audience were Hilbert and another famous
mathematician, Felix Klein. When they realised how interesting and
important the problem was, they both started their own research.
Moreover, Einstein and Hilbert began to discuss their experiences, and in
the ensuing months they frequently exchanged letters informing each
other about current results of their research.[150]

In November 1915, Einstein decided that he should return to the
matter of the general covariance of the gravitational field equations. On
November 4, at the plenary session of the Prussian Academy of Sciences
he presented his new version of the theory of gravitation, titled "Toward
the General Theory of Relativity"[151] and a week later he added an
Appendix.[152] In the introduction to his new article he wrote:

> Thus I returned to the need for a more general covariance of the field
> equations, from which I had reluctantly distanced myself three years
> before, when I worked with my friend Grossmann.[153] **c29**

This "more general covariance" did not constitute the full general
covariance because it was limited to the class of uni-modular
transformations (*i.e.* determinant equal to one).[154] In his Appendix,
Einstein gives another new limiting condition. As a result, neither paper
constituted a fully satisfactory formulation of the General Theory of
Relativity, although both correctly delineated certain phenomena.

Einstein felt that both works led to definite conclusions as regards
the real nature of time and space. In his view, the new theory deprived
space and time of their last remnant of objective reality. He stated this in
the introduction to his new article, "Explanation of the Movement of
Mercury's Perihelion on the Basis of the General Theory of Relativity:"[155]

---

[149] J. Mehra, *Einstein, Hilbert and the Theory of General Relativity* (Dordrecht-Boston: Reidel, 1974), quoted in: M.Heller, *op. cit.*, pp. 16-17 and 21.

[150] M. Heller, "Jak Einstein...," p. 13.

[151] A. Einstein, "Zur allgemeinen Relativitätstheorie," *SPAW* (1915), 2. Teil, pp. 778-786.

[152] A. Einstein, "Zur allgemeinen Relativitätstheorie, Nachtrag," *SPAW* (1915), 2. Teil, pp. 799-801.

[153] *Ibid.*, p. 799

[154] M. Heller, "Jak Einstein...," p. 13.

[155] A.Einstein, "Erklarung der Perihelbewegung des Merkur aus der allgemeinen Relativitätstheorie," *SPAW* (1915), 2. Teil, pp. 831-839.

In a paper which was recently published in this journal, I have formulated the field equations of gravitation, which are covariant with respect to arbitrary transformations with determinant 1. In an appendix I have shown that these field equations behave in a generally covariant way if the scalar of the energy tensor of "matter" vanishes. I have also shown that there are no objections of principle against the introduction of this hypothesis, by which time and space are deprived of the last trace of objective reality. In the article I am presenting now I have found an important confirmation of the most radical theory of relativity. [156] **c30**

The confirmation to which Einstein referred was a correct calculation of the precession of Mercury's perihelion in agreement with observations, and twice the amount of bending of light rays in a gravitational field (compared with Einstein's previous forecasts), which was later confirmed in 1919.

During the subsequent Thursday session of the Prussian Academy of Sciences on November 25, 1915, in a brief paper titled "The Field Equations of Gravitation"[157] Einstein presented the much anticipated general covariant equations of the gravitational field, which took the following form for the empty space-time continuum:

$$G_{\mu\nu} = 0$$

and for a space-time continuum with matter present:

$$G_{\mu\nu} = -\kappa \left( T_{\mu\nu} - \tfrac{1}{2} g_{\mu\nu} T \right)$$

where $G_{\mu\nu}$ is Ricci's tensor, $T_{\mu\nu}$ is the energy tensor of "matter" (*Energietensor der "Materie"*), $T$ is the scalar of the tensor and $\kappa$ is a constant related to the Newtonian gravitational constant.

Einstein achieved this result by giving up the hypothesis that the equations become covariant if the scalar of the energy tensor of "matter" disappears, and introducing this tensor in a different way. Abandoning this hypothesis, with which he associated "robbing space and time of their last trace of objective reality," did not change his epistemological standpoint. He was still convinced that the new theory deprived space and time of objective reality. He did not change his convictions, because this hypothesis was not responsible for this result; rather, general covariance was to be ensured by this hypothesis. General covariance,

---

[156] *Ibid.*, p. 831.

[157] A. Einstein, "Feldgleichungen der Gravitation," *SPAW* (1915), 2. Teil, pp. 844-847.

even in its final form, has the function of depriving co-ordinate systems of physical meaning, thus leading to the epistemological conclusion. The fact that co-ordinate systems were thereby deprived of physical meaning was stressed by Einstein at the end of his article.

> The postulate of relativity in its most general formulation, which transforms the coordinates of space and time into physically insignificant parameters, of necessity leads to a completely well defined theory of gravitation.[158] **c31**

The final formulation of the General Theory of Relativity was presented in detail in the article "The Foundation of the General Theory of Relativity,"[159] which was published at the beginning of 1916. The article—like many of Einstein's other works—contained much interpretation, which was still permeated by Mach's epistemological ideas. Einstein was convinced that his final formulation of the General Theory of Relativity implemented Mach's programme, wherein the notion of real space and time should be removed from physics as an unnecessary metaphysical interpolation. He also believed that it confirmed Mach's principle. The postulate of general covariance, as a consequence, deprived space and time of the reality erroneously attributed to them. Only the coincidence of events was considered to be truly real.

> That this requirement of general covariance, which takes away from space and time the last remnant of physical objectivity, is a natural one, will be seen from the following reflexion. All our space-time verifications invariably amount to a determination of space-time coincidences.[160] **c32**

Is it really true that equalisation of all co-ordinate systems—which according to the new theory lose all physical meaning and are only utilised as a tool for description—demanded denial of real time and space existence? Isn't such a conclusion somewhat exaggerated, if not unfair? Could the General Theory of Relativity not be reconciled with the existence of real space and time, or even better—a real space-time continuum? Einstein was not consistent in his "Foundation" article. On the one hand, he spoke of depriving space and time of the last remnant of reality; on the other hand, he spoke of a four-dimensional space-time

---

[158] *Ibid.*, p. 847.
[159] A. Einstein, "Die Grundlage der allgemeinen Relativitätstheorie," *AdP* (49), 1916, pp. 769-822.
[160] *Ibid.*, p. 774.

continuum, or a four-dimensional metric space, the metric behaviour of which was described by the tensor $g_{\mu\nu}$ defining the gravitational field.

> [...] the ten functions [$g_{\mu\nu}$] representing the gravitational field at the same time define the metrical properties of the [four dimensional] space measured.[161] **c33**

When he wrote this, Einstein failed to notice that he was already linking the notion of space with the carrier of metric structure, thus introducing an ultra-referential real space-time continuum which, according to his theory, had real physical attributes. He was to notice this several months later, in June 1916, in an exchange of letters with Lorentz.

## 2.12  Summary

Summing up this chapter, we note once again that Einstein's concepts of space and time were closely linked to the concepts of reference and co-ordinate systems. Therefore, elimination of the privileged reference system (or absolute space, in Einstein's view), as was accomplished by the Special Theory of Relativity, and elimination of the privileged class of inertial systems, accomplished by the General Theory of Relativity, were treated by Einstein (due to the influence of positivistic philosophy) as the elimination of the concepts of space and time from physics. For Einstein, following Drude's notion, the ether was nothing but space with physical attributes. Consequently, when he concluded that his General Theory of Relativity had deprived space and time of the last remnant of reality, he was even more convinced that the ether had no further place in physics. As we indicated earlier, Einstein would radically change his opinions in a few months. This transformation in Einstein's ideas will be discussed in the next chapter.

---

[161] *Ibid.*, p. 777.

# Chapter 3

# EINSTEIN INTRODUCES HIS NEW CONCEPT OF THE ETHER (1916-1924)

## 3.1    Correspondence with Lorentz, polemic with Lenard

It often happens that new ideas and concepts are born and mature through discussions and polemics. Einstein's new concept of the ether was born out of an exchange of letters with Lorentz and his polemics with Lenard. We could even say that Einstein was provoked to introduce, and stimulated to develop, a new relativistic concept of the ether by these two physicists. Hermann Weyl, who in 1917 presented a version of the relativistic ether similar to Einstein's, may have inspired him to some extent.

In June 1916, Lorentz wrote Einstein a long, article-like letter, and this would provoke Einstein to introduce a new concept of the ether. In order to understand Lorentz's reasons for writing the letter we must recapitulate the events of Lorentz's academic career and scientific achievements. At the end of the nineteenth and at the beginning of the twentieth century, Lorentz tried to formulate his own theory of gravitation.[162] Following Mossotti—who imagined matter as a substance composed of positive and negative electricity, and gravitation as a "partial

---

[162] H. A. Lorentz, *Proc. Acad. Amsterdam*, 2 (1900), p. 559, note taken from H. A. Lorentz, "La gravitation," *Scientia*, 16 (1914), pp. 28-59.

force" resulting from a minute difference between the attraction forces of electricity of reverse sign, and repulsive forces of electricity of the same sign—Lorentz tried to develop these ideas within the science of electromagnetism he had formulated himself. In the early years of this century it seemed to him that the results of this development had proved satisfactory. The year 1906 was a turning point, as Lorentz learned of the relativity principle formulated by Einstein. Because he concluded that the theory of gravitation he developed contradicted the principle, which he felt had to be respected, Lorentz decided that he should abandon the electromagnetic theory of gravitation, and concern himself more with some kind of relativistic theory. From that time on he kept abreast of the works of relativists such as Poincaré, Minkowski, Willem de Sitter, and Einstein, as we are told in his paper "Gravitation."[163] Most of all, he appreciated Einstein's attempts to formulate a new theory of gravitation. He was particularly interested and committed to them, writing many articles and lecturing on the subject. He also conducted a lively letter exchange with Einstein, so we can even say that Lorentz contributed to the development of the new theory. The details of his contribution and contacts with Einstein can be found in J. Illy's article "Einstein Teaches Lorentz, Lorentz Teaches Einstein. Their Collaboration in General Relativity."[164]

The common feature of Lorentz's publications and lectures on the developing General Theory of Relativity was his persistent efforts to reconcile Einstein's theory of gravitation with his model of a stationary ether. His "Gravitation" may serve as an example; it was written after Einstein and Grossmann had published their first joint article. On the one hand, Lorentz was full of enthusiasm about the achievements of Einstein and Grossmann, as he admitted with full conviction that the gravitational field was described by metric tensor $g_{\mu\nu}$; on the other hand, however, he tried to prove that these new achievements could be reconciled with the concept of a stationary ether.[165]

Lorentz greeted the final formulation of the General Theory of Relativity with satisfaction and appreciation. However, he still claimed that the General Theory of Relativity could be reconciled with the concept of an ether at rest. He tried to convince Einstein of this, and this

---

[163] H. A. Lorentz, "La gravitation," pp. 28-59.

[164] J. Illy, "Einstein Teaches Lorentz, Lorentz Teaches Einstein. Their Collaboration in General Relativity, 1913-1920," *AHESc*, 39 (1989), pp. 247-289.

[165] H. A. Lorentz, "La gravitation," p. 58.

is why he wrote his long letter of June 6, 1916. The following excerpt is representative of the kind of argument Lorentz used:

> During the past few months I have spent considerable time with your theory of gravitation and the general theory of relativity, and I lectured on it, which was very useful for me. I believe now that I understand the theory in its full beauty, as every difficulty I encountered could be overcome by reflecting on it further. I also succeeded in deducing your field equations
>
> $$G_{\mu\nu} = -\kappa \left( T_{\mu\nu} - \tfrac{1}{2} g_{\mu\nu} T \right)$$
>
> from the variational principle, or at least only a detail is missing in this deduction, which required long calculations.
>
> I have now arrived at a notion, which I would like to present to you, which is based on the consideration of a fictitious experiment. Imagine performing Lecher's experiment with two perfectly conducting wires, which extend around the earth at the equator, each one being closed on itself. In order not to risk "derailing" the electromagnetic waves (caused by the earth's curvature) we can use one wire with a concentrical conducting sleeve, instead of the two wires. At a certain point, $A$, of this "cable"** closed upon itself, there should be a device which makes it produce waves, and a detector with which we observe the returning waves at $A$ as they complete the circle. The cable and the point $A$ should be fixed to the earth.
>
> Based on what we know we could accurately predict what we would observe with sufficiently sensitive instruments. Waves which are produced at the same time at $A$ and which cross the circumference in opposite directions will *not* return to $A$ at the same time.
>
> Among the different ways in which we can describe this result, there are only two which are particularly simple.
>
> **a.** We can choose a system of coordinates I OX, OY (OZ must coincide with the axis of the earth) in such a way that in this system the transmission velocity of the waves for both crossing directions is the same. We find then that the earth rotates in this system of coordinates.
>
> **b.** We introduce a system of coordinates II, which is firmly fixed to the earth. With respect to it, there are different propagation velocities $c_1$ and $c_2$ for the two circulation directions.
>
> It is almost superfluous to say that the necessary difference of the propagation velocities results from your general formulas, when passing from I to II, and to the extent that an equation of the form

$$c_1 - c_2 = a$$

holds in system I as well as in II, and also in many other systems (every time with a different *a*), we can say that it expresses the result of the experiment in a *covariant* form. But this must not deter us from considering the equality $c_1 - c_2 = 0$ as *different* from $c_1 - c_2 \neq 0$. In this sense we will conclude: the propagation in the cable does not behave in the same way with respect to the systems of coordinates I and II.

Now, if we try to make this somehow understandable, and to represent it pictorially, it will be hardly possible to speak *only* of the earth, of the cable and of the "space" or "vacuum" contained in the latter. It will be really impossible to imagine that nothing exists in space or in vacuum that behaves differently with respect to the systems I and II.

The representation is very easy to find [I will speak of another one below] and was previously considered to be very natural by all physicists: in the cable there is a medium (ether) *in which* the waves propagate, such that the propagation velocity relative to the medium is always the same; this medium is at rest if referred to one system of axes, and it can be in motion if referred to another system of axes. If we start from this point of view we can say that the experiment has shown us the motion of the earth relative to the ether. Thus, having recognised the possibility of detecting relative *rotation*, we cannot *a priori* deny the possibility of obtaining the effects of a *translation* of the same kind as well; that is to say, we should not make the starting assumption of the theory of relativity a postulate. We must instead search for the answer to the question in the observations (and this was also the real path of development). Having learned from them that an influence from translation cannot be found, then generalising (and certainly in a broad way), we can express that principle as a fundamental *hypothesis*, which still allows the possibility (however improbable it may be seem to us) that careful observation will force us to drop the hypothesis in the future.

These considerations could also be expressed in another way. We can produce standing waves in the closed cable and observe the position of the nodes at every instant. It will then be found that the waves propagate around the earth in a form of a circle. We could then simply observe the relative motion of the nodes with respect to the earth (or *vice versa*). However, if we consider that the same rotation appears in standing waves of different length and different intensity, it becomes natural (let us say as a vivid summary of all these phenomena) to think of an ether in which the standing waves are based.

In his discussion of a similar experiment Mach, whose conception you have followed, felt the need to accept something lying outside the earth which would determine these phenomena. In his way of thinking one would look for a determining moment in the influence of the "distant bodies of the universe," let us say the fixed stars. One might say that the fixed stars moving around in a circle (or at rest) determine the nodes in the ring shaped cable. Although this conception seems to me much less reasonable than the hypothesis of an ether, I could accept it if it offered an advantage of some kind in comparison with this hypothesis. But I am unable to see any. If, however, we must accept that the rotation of the earth with respect to the fixed stars has an observable influence on electromagnetic phenomena, we cannot deny *a priori* the possibility of an analogous influence by the *translation* of the earth or of the solar system relative to the fixed stars. We arrive then at exactly the same point as with the ether hypothesis, and we must carry out experimental research to find if there exists some consequence of a translation. Now there would be no reason for a relativity *postulate*.

Of course, the two points of view—influence of the fixed stars and the ether hypothesis—both come into play, and in my opinion they are not very different from one another. Suppose that I assume that the motion or state of rest of the nodes in our ring shaped cable is determined by the influence of the fixed stars. Then, in order to approximately define the nature of this influence, I can assume that inside the cable there is a system of rigidly connected points linking the fixed stars and the electromagnetic waves. I could say that the influence in question manifests itself in the fact that the nodes have stable positions with respect to this system of points, which is also linked to the fixed stars. Going from this system of points to an ether is not a big step.

Of course there are also other possibilities, for example those discussed by you and Mach with very similar considerations, and the previous reflections will in no way be new to you. The main point is, indeed, that deviations from the theory of relativity would also be quite plausible within the "fixed stars hypothesis." I do not need to say that both the theory of relativity and your gravitational theory can also remain fully valid under the point of view I have proposed, except that they will no longer appear as the only possibility.[166] **d1**

(Note by Lorentz: **space between conductors empty of air).

---

[166] H. A. Lorentz, "Letter to A. Einstein, 6/6/1916," *EA*, 16-451.

Einstein answered Lorentz's letter promptly. His answer was dated 16 June 1916. Einstein, naturally, did not agree with the challenge to his postulate or the relativity principle. Nor did he agree with Lorentz's opinion that the General Theory of Relativity could be reconciled with the concept of the stationary ether. For the first time, however, there emerged a concept of a new, non-stationary ether which would not violate the relativity principle. He noted that the space-time continuum described in his General Theory of Relativity was something real, characterised by physical attributes, and therefore he placed an equals sign between the space-time continuum of the General Theory of Relativity, the state (metric behaviour) of which is described by the metric tensor $g_{\mu\nu}$ and the ether. A. Miller first noticed the concept of the new ether in this very letter.[167]

Here we quote the excerpt from Einstein's letter in which he analysed Lorentz's arguments and wrote of the possible introduction of a new theory of the ether:

> Let us examine your thoughts on interference now! I was amused that you have considered exactly the same example that went through my head often in the past few years. I agree with you that the general theory of relativity is closer to the ether hypothesis than the special theory. This new ether theory, however, would not violate the principle of relativity, because the state of this $g_{\mu\nu}$ = ether would not be that of a rigid body in an independent state of motion, but every state of motion would be a function of position determined by material processes. Example:

---

[167] A. Miller, *Imagery in Scientific Thought Creating 20th Century Physics* (Boston-Stuttgart: Birkhauser, 1984), p. 55.

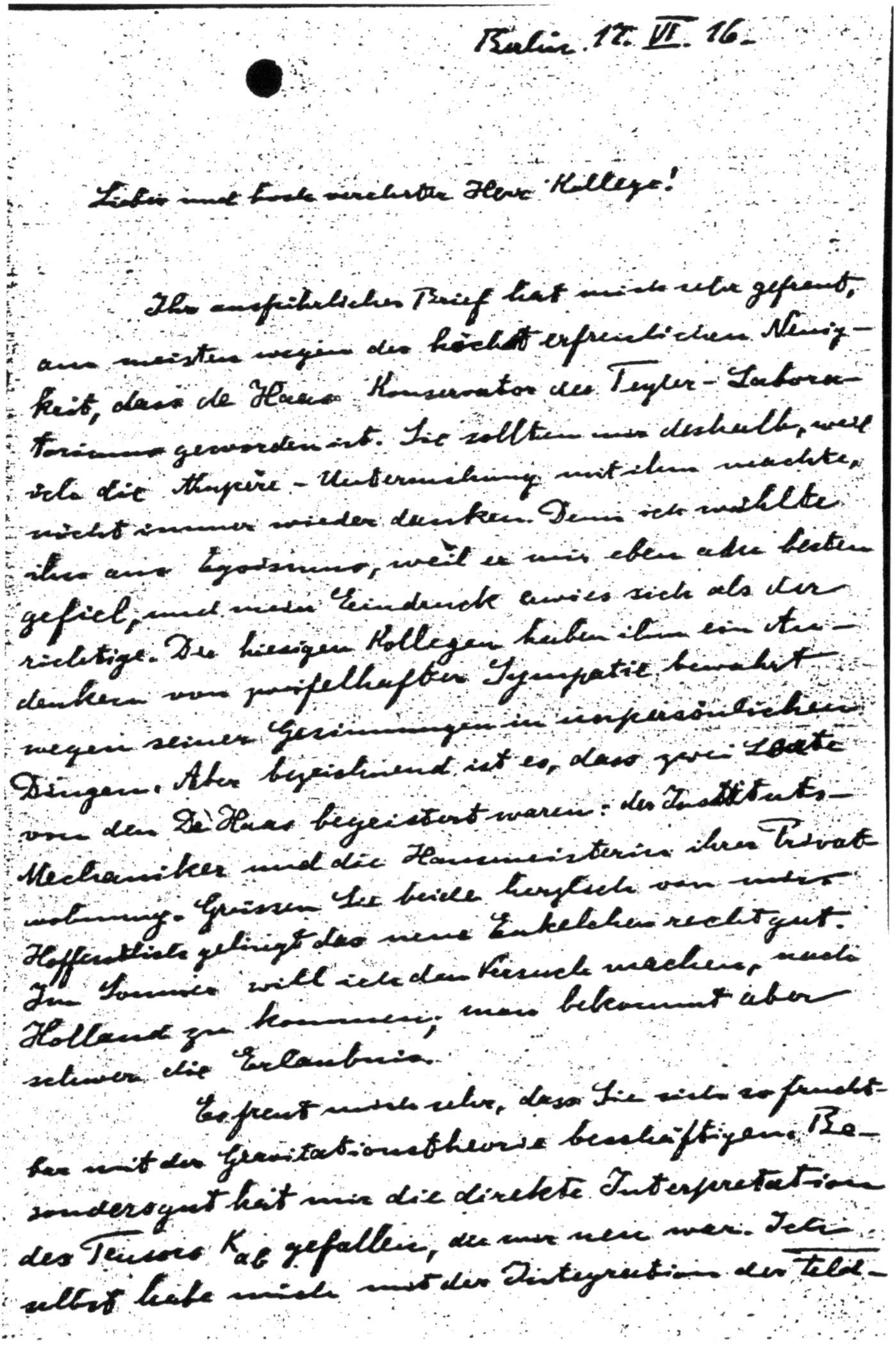

Berlin 17. VI. 16.

Lieber und hochverehrter Herr Kollege!

Ihr ausführlicher Brief hat mich sehr gefreut, am meisten wegen der höchst erfreulichen Neuigkeit, dass de Haas Konservator des Teyler-Laboratoriums geworden ist. Sie sollten nur deshalb, weil ich die Theorie-Untersuchung mit ihm machte, nicht immer wieder danken. Denn ich wählte ihn aus Egoismus, weil er mir eben am besten gefiel, und mein Eindruck erwies sich als der richtige. Die hiesigen Kollegen haben ihm ein Andenken von zweifelhafter Sympathie bewahrt wegen seiner Gesinnungen in unpersönlichen Dingen. Aber bezeichnend ist es, dass zwei Leute von den De Haas begeistert waren: der Institutsmechaniker und die Hausmeisterin ihrer Privatwohnung. Grüssen Sie beide herzlich von mir. Hoffentlich gelingt das neue Enkelchen recht gut. Im Sommer will ich den Versuch machen, nach Holland zu kommen; man bekommt aber schwer die Erlaubnis.

Es freut mich sehr, dass Sie sich so fruchtbar mit der Gravitationstheorie beschäftigen. Besonders gut hat mir die direkte Interpretation des Tensors $K_{ab}$ gefallen, die mir neu war. Ich selbst habe mich mit der Integration der Feld-

**Figure 2a-d**. The text of Einstein's letter to H. A. Lorentz dated June 17, 1916, in which Einstein introduces the notion of the new ether for the first time. (Reproduced with the kind permission of The Albert Einstein Archives, The Jewish National and University Library, The Hebrew University of Jerusalem, Israel.)

gleichungen in erster Näherung berechnet und die Gravitationswellen untersucht. Die Resultate sind zum Teil überraschend. Es gibt dreierlei Wellen, von denen aber nur eine typus Energie transportiert. Mit der Theorie der Ausstrahlung materieller Systeme bin ich noch nicht ganz fertig. Aber soviel ist mir klar, dass die Quanten-Schwierigkeiten auch die neue Gravitationstheorie treffen, ebenso gut wie die Maxwell'sche Theorie. Es hat mich sehr gefreut, dass Sie in Ihrem Pariser Vortrage die Schwankungs-Eigenschaften der Strahlung einer eingehenden Behandlung gewürdigt haben, hier treten die Unverträglichkeiten der Theorien am reinsten zutage.

Nun zu Ihrer Interferenzbetrachtung. Es hat mich amüsiert, dass Sie genau auf dasselbe Beispiel verfallen sind, das auch ich mir in den letzten Jahren habe oft durch den Kopf gehen lassen. Ich gebe Ihnen zu, dass die allgemeine Relativitätstheorie der Aetherhypothese näher liegt als die spezielle Relativitätstheorie. Aber diese neue Aethertheorie würde das Relativitätsprinzip nicht mehr verletzen. Denn der Zustand dieses $g_{\mu\nu}$=Aethers wäre nicht der eines starren Körpers von selbständigem Bewegungszustande.

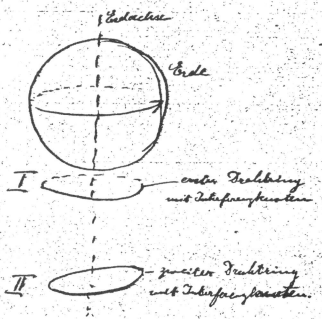

Massgabe der Verdrehung und der Entfernungen. Das Foucaultsche Pendelleben dreht sich auch ein wenig mit der Erde, etwa 0,01" per Jahr. Schade, dass es nicht mehr ausmacht! Ich muss aber gestehen, dass mir das ganze System lieber ist als ein unvollkommener Vergleich mit einem stofflichen Etwas. Denn die Bevorzugung der gleichförmigen Bewegung findet in dieser modifizierten Äther-hypothese keinen Ausdruck, wohl aber im abstrakten System. Geht man nämlich von einem Weltstück von Konstanten $q_{\mu\nu}$ aus, so ändert eine lineare Substitution der $x_\nu$ nichts an der Konstanz der $q_{\mu\nu}$, wohl aber eine nicht-lineare Substitution der $x_\nu$. Hieraus folgt, dass gleichförmige Relativbewegung kein Gravitationsfeld erzeugt d. h. unmerklich ist im Gegensatz zur ungleichförmigen Bewegung. Jener fundamentale Unterschied von gleichförmig und ungleichförmig kommt aber in der Äthervorstellung nicht unmittelbar zum Ausdruck; man möchte vielmehr stets eine gleichförmige Bewegung nachweisen können.

    Mit herzlichen Grüssen und den besten Wünschen für die Gesundheit Ihrer Frau

            Ihr

                A. Einstein.

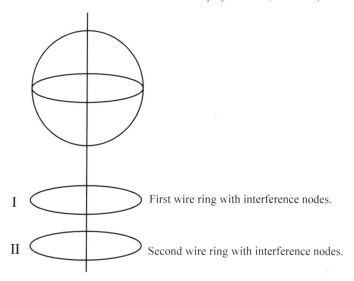

I — First wire ring with interference nodes.

II — Second wire ring with interference nodes.

If the earth did not exist or did not rotate, the interference nodes of ring I and II would remain at rest in relation to the "fixed stars" and also in relation to each other. But the earth rotates and both node systems rotate together with it, even if in a very tiny percentage, and specifically the system of nodes of I, because of the smaller distance, more than that of II. The systems of nodes I and II rotate also with respect to one another with a very small velocity determined by the amount of the earth's rotation and the distances. Also the plane of Foucault's pendulum rotates a little together with the earth, about 0.01" per year. Unfortunately it does not rotate more. I must however admit, that I prefer the system of $g_{\mu\nu}$ to an incomplete analogy with a material something. This is because the privileged nature of the uniform motion does not find any expression in these modified hypotheses of an ether, while it does in the abstract system. In fact, if one starts from a region of the world with constant $g_{\mu\nu}$, a linear substitution of the $x_\nu$ does not change the constancy of $g_{\mu\nu}$, while a nonlinear substitution of the $x_\nu$ certainly does so. From this it follows that uniform relative motion does not "produce" any gravitational field, that is to say, it remains unnoticed, contrary to non uniform motion. This fundamental difference between uniform and non uniform is, however, not expressed directly in the description with an ether. One would much prefer always to be able to specify a uniform motion.[168] **d2**

---

[168] A. Einstein, "Letter to H. A. Lorentz, 17/6/ 1916," *EA*, 16-453.

As we can see, Einstein once again rejected the existence of the stationary ether, which violated his relativity principle, *i.e.* he rejected the ether understood as a rigid material medium which has its own reference system in which it is at rest. According to Einstein, a new ether could be introduced instead: it would not violate the relativity principle because it would not be a medium like a rigid body which had its own state of motion, and which would determine the state of the movement of particles similarly to the field defined by material processes, like the presence and motion of matter. The state of the new ether (its metric behaviour) would be described by the $g_{\mu\nu}$-tensor and that is why Einstein wrote "*state of the $g_{\mu\nu}$ = ether.*"

The new ideas were not ready for publication, partly because of Einstein's state of mind. He was too deeply engaged in fighting the old ether to be willing to introduce a new one. When we read the letter quoted above, we cannot say that he was enthusiastic about spreading his new ideas. He only admitted that the General Theory of Relativity was closer to the notion of the ether than the Special Theory of Relativity, and that if he could introduce an ether that did not violate the relativity principle, it would have to be identified with this reality, which was locally described by the metric tensor $g_{\mu\nu}$. For the next two years after he wrote the letter to Lorentz, Einstein did not include a single hint of the possibility of introducing a new hypothesis of the ether in any of his articles on the theory of relativity.

In the middle of 1918 a book of lectures on the theory of relativity (chiefly the General Theory of Relativity) held in the summer semester in 1917 at the Technical University of Zürich was published by the renowned mathematician Herman Weyl. Weyl hit upon an idea similar to Einstein's idea of treating the reality described by the metric tensor $g_{\mu\nu}$ as the new ether. The following is an excerpt from his book:

> The coefficients of the fundamental metric form [components of the tensor $g_{\mu\nu}$—L.K.] are therefore not simply the potentials of the gravitational and centrifugal forces, but *determine in general, which points of the universe are in reciprocal interaction.* For this reason the name "gravitational field" is perhaps too unilateral for the reality described by this expression and should better be replaced by the word "ether"; while the electromagnetic field should simply be called field. In reality this "ether" plays the same role as the ether of the old theory of light and of "absolute space" of Newtonian mechanics; only one must not

forget, that it is something completely different from a substantial carrier.[169] **d3**

The idea of treating the reality described by the metric tensor $g_{\mu\nu}$ as the ether was quite original here, as he failed to mention Einstein's name and wrote as if the concept was his, not Einstein's.

Einstein readily accepted Weyl's book and wrote a very positive review of it.[170] Perhaps Weyl's publishing of the concept of the relativistic ether pushed Einstein, to some extent, to publish his ideas. The major reason, however, was Lenard's attack upon the theory of relativity.

In July 1917 Lenard delivered a paper titled "Relativity Principle, Ether, Gravitation," which was at first published in the form of an article[171] and then, in wider circulation, as a brochure.[172] In his text he attacked the relativity principle and, without any mathematical support, explained gravitation by a specific kind of composite ether. He claimed the General Theory of Relativity simply renamed[173] the ether "space," or—in other words—that the General Theory of Relativity could not exist without the ether. Although it denied the ether's existence, the ether re-appeared in the theory under a different name.

> It is remarkable that precisely the generalised principle of relativity, which seems to oppose ether with a particular force of exclusion, arrives at "space coordinates," which are essentially characteristic of this principle, but which—owing to the variability of their properties—may be well suited to define the states of space. Hence, one gets even the impression that precisely that ether which had been excluded now reappears here with the its name changed to "space."[174] **d4**

In reply to Lenard's accusations, Einstein published his new concept of the ether in his "Dialogue concerning Accusations against Relativity Theory" in November 1918.[175] The article was written in the form of a

---

[169] H. Weyl, *Raum, Zeit, Materie* (Berlin: J. Springer, 1918).

[170] A. Einstein, [review.] "H. Weyl, Raum, Zeit, Materie," *Die Naturwissenschaften*, 6 (1918), p. 373.

[171] Ph. Lenard, "Über Relativitätsprinzip, Äther, Gravitation," *JR*, 15 (1918), pp. 117-136.

[172] Ph. Lenard, *Über Relativitätsprinzip, Äther, Gravitation* (Leipzig: S. Hirzel, 1918), new editions: 1920, 1921.

[173] *Ibid.*, the 1920 edition, p. 7.

[174] *Ibid.*, p. 28.

[175] A. Einstein, "Dialog über Einwande gegen die Relativitätstheorie," *Die Naturwissenschaften*, 6 (1918), pp. 697-702.

dialogue between the Critic, mentioned by name (*i.e.* Lenard) several times, and the Relativist who represents Einstein's point of view. What follows is an excerpt from the dialogue that refers to the ether:

> The Critic: So how is life now for that sick man of theoretical physics, the ether, which most of you had declared irrevocably dead?

> The Relativist: He has had a changing destiny, and it is impossible to say that he is dead now. Before Lorentz he existed as an all penetrating fluid, as a gaseous fluid or in other very different forms of existence, changing from author to author. With Lorentz he became rigid and impersonated a system of coordinates "at rest," that is to say a privileged state of motion in the Universe. According to the special theory of relativity a privileged state of motion did not exist anymore; this meant the negation of ether in the sense of earlier theories. Because if an ether existed it should have had a specific state of motion at every space-time point, which should have played a role in optics. But such a privileged state of motion does not exist, as the special theory of relativity teaches, and therefore also an ether in the traditional sense does not exist. Nor does the general theory of relativity recognise a privileged state of motion at any point, which could somehow be interpreted as the velocity of an ether. Whereas according to the special theory of relativity a part of space without matter and without electromagnetic field seems to be completely empty, that is to say not characterised by any physical properties, according to the general theory of relativity even space that is empty in this sense has physical properties. These are characterised mathematically by the components of the gravitational potential [$g_{\mu\nu}$ tensor—L.K.], which describe the metric behaviour of this part of space, as well as its gravitational field. This state of things can be easily understood by speaking about an ether, whose state varies continuously from point to point. One must only be careful not to attribute to this "ether" the properties of ordinary material bodies (*e.g.*, a well defined velocity at every point).[176] **d5**

As we can see, provoked by Lenard's accusations, Einstein published his new concept of the ether. He actually agreed with Lenard that the General Theory of Relativity's space had physical properties. He did not agree, however, to the return of the old ether, which was in a definite state of motion. This ether is once again renounced as non-existent. Encouraged by Lenard's brochure, another German physicist, Ernst Gehrcke, published an article in which he gave grounds for the existence

---

[176] *Ibid.*, pp. 701-702.

of the ether in the old sense of the word.[177] Einstein then published a brief note[178] showing that Gehrcke's ideas were erroneous, and noting mistakes he had made. The controversy between Lenard and Gehrcke on the one hand, and Einstein on the other soon turned into an anti-Einstein campaign with a strong anti-Semitic flavour. Lenard and Gehrcke, together with other German physicists then began to create a "German" physics in opposition to "Jewish" physics.

## 3.2    The new ether concept in the *"Morgan Manuscript"*

In the meantime, Einstein had become famous in England, his fame being connected with the success of the scientific expedition organised by an English physicist, Arthur Eddington, which confirmed the bending of stellar rays in the vicinity of the Sun. *Nature* magazine decided to print a special edition devoted to the theory of relativity only. Einstein was invited to write a survey which would present the development of the theory. He wrote a thirty-five page article in which two paragraphs were devoted to the new concept of the ether. He finished writing the article in January 1920.[179] Since the article was too long, only a *précis* was published under the title "A Brief Outline of the Development of the Theory of Relativity,"[180] which did not include the paragraphs on the new ether. The manuscript of the whole article, titled "Chief Notions and Methods of the Theory of Relativity Presented in its Development,"[181] has been preserved, and it is kept in the Morgan Library in New York. As a result, it is often referred to as the *"Morgan Manuscript."* We reproduce the two paragraphs devoted to the ether here:

### (13) Special relativity and the ether.

It is clear that in the theory of relativity there is no place for the notion of ether at rest. If the reference systems $K$ and $K'$ are completely equivalent for the formulation of the laws of nature, it is inconsistent to base the theory on a conception that distinguishes one of these systems from the others. If one postulates an ether at rest with respect to $K$, it moves with respect to $K'$, which is not in accord with the equivalence of the two systems.

---

[177] E. Gehrcke, "Über den Äther," *VDPG*, 20, 1918, pp. 165-169.

[178] A. Einstein, "Bemerkung zu E. Gehrckes Notiz: Über den Äther," *VDPG*, 20 (1918), p. 261.

[179] A. Pais, *op. cit.*, p. 177.

[180] A. Einstein, "A brief outline of the development of the theory of relativity," *Nature*, 106 (1921), pp. 782-784.

[181] A. Einstein, (*Morgan Manuscript*) *EA*, 2070.

Therefore, in 1905, I was of the opinion that it was no longer allowed to speak about the ether in physics. This opinion, however, was too radical, as we will see later when we discuss the general theory of relativity. It is still permissible, as before, to introduce a medium filling all space and to assume that the electromagnetic fields (and matter as well) are its states. But, it is not permitted to attribute to this medium a state of motion at each point, by analogy with ponderable matter. This ether may not be conceived as consisting of particles that can be individually tracked in time.[182] **d6**

## (22) General relativity and ether.

It is not difficult to incorporate the laws of nature, already known from special relativity, into the broader framework of general relativity. The mathematical methods were readily available in the "absolute differential calculus," based on the work of Gauss and Riemann and further developed by Ricci and Levi-Civita in particular. It represents a rather simple way of generalising the equations from the special case of the constant $g_{\mu\nu}$ to the case of the spatio-temporary varying $g_{\mu\nu}$. In all laws generalised in this way, a role is played by the gravitational potentials $g_{\mu\nu}$ which, in a word, express the physical properties of empty space.

Thus, once again "empty" space appears as endowed with physical properties, *i.e.*, no longer as physically empty, as seemed to be the case according to special relativity. One can thus say that the ether is resurrected in the general theory of relativity, though in a more sublimated form. The ether of the general theory of relativity differs from the one of earlier optics by the fact that it is not matter in the sense of mechanics. Not even the concept of motion can be applied to it. It is furthermore not at all homogeneous, and its state has no autonomous existence but depends on the field-generating matter. Since in the new theory, metric facts can no longer be separated from "true" physical facts, the concepts of "space" and "ether" merge together. Since the properties of space appear as determined by matter, according to the new theory, space is no longer a precondition for matter; the theory of space (geometry) and of time can no longer be presupposed prior to actual physics and expounded independently of mechanics and gravitation.[183] **d7**

---

[182] *Ibid.*, section 13.
[183] *Ibid.*, section 22.

## 3.3     The anti-Einstein campaign over the ether

As we have already mentioned, the controversy between Einstein and Lenard together with Gehrcke broke out over the existence of the ether. The dispute with Lenard started as early as 1910. Gehrcke attacked the General Theory of Relativity on the same issue immediately after its publication in his article "On Critics and History of the New Theories of Gravitation."[184] In the first paragraph of his article he proved that Einstein used an example in his paper that Gehrcke had used in his work of 1911.[185] He had discussed a system of two spheres rotating against each other along the same axis; one of them maintained its spherical shape, whereas the other was distorted due to centrifugal forces. Einstein explained this distortion as due to interaction with the distant stars against which the distorted sphere rotated. According to Gehrcke, it was the motion relative to the ether that mattered, and not motion relative to distant masses.

In paragraph 2 Gehrcke tried to show that Einstein's General Theory of Relativity could not be accepted. He particularly disliked the equivalence between uniformly accelerated motion and a homogenous gravitational field. One of his reasons for rejecting the general relativity principle was that it could not be reconciled with the existence of the ether.

In paragraph 3 Gehrcke accused Einstein of plagiarism, as he proved that his mathematical formula describing the shift in Mercury's perihelion was identical with a formula published by Paul Gerber as early as 1898.[186] According to Gehrcke, Einstein knew Gerber's equation very well, having read Mach's *Mechanics*. This was not true, as Gerber's equation does not appear in Mach's *Mechanics*. Mach merely mentioned Gerber's achievements without quoting his mathematical formula.[187]

Gehrcke attacked the theory of relativity without showing any mercy for its creator. He also actively joined an anti-Einstein campaign with strong political and anti-Semitic leanings organised in 1920 by Paul Weyland, who was not a physicist himself, but was a founder of the

---

[184] E. Gehrcke, "Zur Kritik und Geschichte der neueren Gravitationstheorien," *AdP*, 50 (1916), pp. 119-124.

[185] E. Gehrcke, *VDPG*, 13 (1911), pp. 666 and 993. Note taken from: E. Gehrcke, "Zur Kritik und Geschichte...," p. 120.

[186] P. Gerber, "Über die raumliche und zeitliche Ausbreitung der Gravitation," *ZMPh*, 43 (1898), p. 93.

[187] E. Mach, *Die Mechanik in ihrer Entwicklung historisch-kritisch dargestellt*, 5th ed. (Leipzig: Brockhaus, 1904), p. 201.

Association of German Scientists for the Preservation of Pure Science. Einstein's biographer, the renowned physicist Philipp Frank, who took over the Chair of Physics after Einstein left Prague, described Weyland and his campaign in the following way:

> Suddenly an organisation appeared whose sole objective was to fight against Einstein and his theories. Its leader was a certain Paul Weyland, about whose past, qualifications and occupation nobody knew anything. The organisation had a lot of money of unknown origin. It offered relatively high honoraria to people who wanted to write or speak against Einstein in public meetings. It held assemblies in the largest concert hall in Berlin and publicised them with gigantic posters, of the type usually reserved only for performances by the most famous virtuosi.[188] **d8**

Weyland would read from the works of Lenard and Gehrcke criticising the theory of relativity, and fished out everything that might serve his purpose. Before organising lectures at the Berlin Philharmonic Concert Hall he started a press campaign which began by quoting Lenard and Gehrcke. In an article published in *Tägliche Rundschau* of August 6, 1920, titled "Einstein's Theory of Relativity as Scientific Mass Suggestion,"[189] he quoted Lenard's *Relativity Principle, Ether, Gravitation* to attack the generalised relativity principle, which excluded the ether in the old sense of the word. He also repeated Lenard's accusation that Einstein had renamed the ether "space":

> That Einstein eliminated the ether by decree, that he re-introduced it via a different concept with the same functions, should be mentioned for its "drollery," to speak like Einstein.[190] **d9**

Seeking to discredit Einstein, Weyland repeated Gehrcke's accusation of plagiarism. Gehrcke, however, had been rather subtle, whereas Weyland did it in a vulgar manner.

Max von Laue, a well-known physicist, defended Einstein, answering Weyland's article in the same newspaper on August 11, 1920.[191] He defended Einstein's generalised relativity principle against the attacks by Lenard and Gehrcke with a simple example. He also noted that Einstein's mathematical formula describing the shift in Mercury's perihelion, which

---

[188] Ph. Frank, *Einstein. His Life and Times* (New York: Knopf, 1953), p. 159.

[189] P. Weyland, "Einsteins Relativitätstheorie—eine wissenschaftliche Massensuggestion," *Tägliche Rundschau*, August 6, 1920, Evening Edition.

[190] *Ibid.*

[191] M. von Laue, "Zur Erörterung über die Relativitätstheorie," *Tägliche Rundschau*, August 11, 1920, Evening Edition.

accidentally resembled Gerber's formula, was derived by Einstein from the mathematical formalism of the General Theory of Relativity as one of its consequences, which made it a theoretically sound law. He also mentioned the fact that a Munich astronomer, H. Von Seeliger, who was a declared opponent of the theory of relativity, had found vagueness, inaccuracy and errors in Gerber's assumptions and mathematical operations. Thus—Laue claimed—Gerber's derivations and deliberations should not be put on a par with the General Theory of Relativity and Einstein's derivations.

In reply, Weyland published a brief note on the same day,[192] declaring that he and Gehrcke would take their stand on the issues at an open meeting to be held in the main auditorium of the Berlin Philharmonic Concert Hall on August 24, 1920. He invited Laue to the meeting.

Laue did not accept the invitation. Einstein, however, arrived at the meeting, together with his daughter.[193] Many representatives of Berlin's academic society also attended, including Professor Walther Nernst. A Polish physicist, Leopold Infeld, who was a student at that time, was there as well.[194] At the entrance, the audience could buy swastikas to pin into a lapel.[195] The meeting started with Weyland's presentation titled "On Einstein's Theory of Relativity and the Way it was Introduced."[196] The next to take the floor was Gehrcke, who gave his lecture the same title Weyland had used for his article, *i.e.* "Einstein's Theory of Relativity as Scientific Mass Suggestion."[197] Einstein, obviously amused by the event, applauded loudly as he was being attacked.[198]

---

[192] P. Weyland, [untitled note following the paper by M. von Laue], *Tägliche Rundschau*, August 11, 1920, Evening Edition.

[193] K. J., "Der Kampf gegen Einstein," *Die Vossische Zeitung*, new edition in: P. Weyland, "Betrachtungen über Einsteins Relativitätstheorie und die Art ihrer Einführung," *Schriften aus dem Verlage der Arbeitsgemeinschaft deutscher Naturforscher zur Erhaltung reiner Wissenschaft and. V.* H. 2. (Berlin: 1920), p. 6.

[194] L. Infeld, *Szkice z przeszłości* (Warszawa: Państwowy Instytut Wydawniczy, 1966), p. 58.

[195] K. J., *op. cit.*, p. 8.

[196] P. Weyland, "Betrachtungen über Einsteins Relativitätstheorie und die Art ihrer Einführung," *Schriften aus dem Verlage der Arbeitsgemeinschaft deutscher Naturforscher zur Erhaltung reiner Wissenschaft and. V.* H. 2. (Berlin: 1920), pp. 10-20.

[197] E. Gehrcke, "Die Relativitätstheorie eine wissenschaftliche Massensuggestion," *Schriften aus dem Verlage der Arbeitsgemeinschaft deutscher Naturforscher zur Erhaltung reiner Wissenschaft and. V.*, H. 1. (Berlin: 1920).

[198] Ph. Frank, *op. cit.*, p. 272.

Weyland started his intervention with a comment that it was an introduction to the series of lectures that would deal with "Einstein's so-called theory of relativity" in order to perform critical research to find out if "Einstein's fictions" could be confirmed by true science. Then he announced that the audience could purchase Gehrcke's lecture and Lenard's brochure in the lounge of the Concert Hall. Einstein naturally realised that Lenard was somehow involved in the whole affair. In the remainder of his speech, Weyland launched an attack against the generalised relativity principle without using any concrete arguments. He only accused it of becoming renowned due to organised propaganda. The supporters of the theory of relativity started to interrupt Weyland's talk, demanding he go into details, though they never heard any concrete arguments. Weyland continued, trying to show that the propaganda he was discussing contributed to the "systematic mass suggestion." Because it had been put to the masses as another Copernican revolution, the General Theory of Relativity had become highly acclaimed. He characterised Einstein's theory as a chaos of thought typical of dadaists. Consequently, one could speak of about "Einstein's phantasms." The closing words of Weyland's speech were another fierce accusation of plagiarism: Einstein had simply copied Gerber's formula and kept silent about it. According to Weyland, many professionals had confirmed the fact. It would be decent, then, for Einstein to break his silence and speak. Einstein did not react, despite the provocation.

In his speech, Gehrcke launched an attack against the relativity principle. It immediately became apparent that he was an adherent of the ether in the old meaning, and a defender of absolute space, so he also attacked the relativisation of space and time, together with the General Theory of Relativity. The existence of the ether, acknowledged—in Gehrcke's opinion—by most eminent scholars past and present, struck a blow at the very heart of the relativity theory, namely at its relativity principle. The existence of the ether broke this principle by distinguishing a single reference system as basic for electrical, magnetic, and optical phenomena. The existence of the ether also threatened the important novelty of the General Theory of Relativity, namely—as he claimed— strange artificial fields of gravitation without any natural source. The forces which emerged in accelerated and rotating systems had their sources in the ether.

At the end of his intervention, Gehrcke returned to the main theme of Weyland's argument: the General Theory of Relativity had been sold to the masses by good advertising, and this was the reason for the title of the lecture.

Einstein was highly amused by the attacks launched against his theory and himself; however, he did not remain silent. Three days later, on August 27, 1920, he published an article "My Response to the Anti-Relativistic Limited Liability Company."[199] In this excerpt he gives his reasons for taking up the challenge:

> I am perfectly aware of the fact that both speakers are not worthy of an answer from my pen; in fact, I have good reason to believe that reasons other than the search of truth lie behind their enterprise. (If I were a German national, with or without a swastika, instead of a Jew with a liberal, international spirit, then...). I answer only because friends have repeatedly expressed the wish that I do so, so that my point of view can be known.[200] **d10**

In another passage he criticises Lenard, whom he believes to have participated in the whole affair. After stating that most world-famous physicists (he mentions the names of Lorentz, Planck, Sommerfeld, Laue, Born, Larmor, Eddington, Debye, Langevin, and Levi-Civita) recognise the cognitive value of the theory of relativity and contributed to it, Einstein goes on to say:

> Among physicists of international reputation I could only mention Lenard as an explicit enemy of the theory of relativity. I admire Lenard as a master of experimental physics; in theoretical physics he has, however, done nothing, as yet, and his objections to the general theory of relativity are so superficial, that until now I did not feel it was necessary to answer them in detail.[201] **d11**

The greater part of Einstein's article was devoted to a criticism of Gehrcke's lecture. Einstein accused Gehrcke of "consciously misleading an incompetent audience" and the accusation of solipsism levelled by Gehrcke in reference to the relativisation of time and space could—in Einstein's opinion—only be treated as a joke.

> Mister Gehrcke declares, that the theory of relativity leads to solipsism, a statement, which every insider will consider a joke.[202] **d12**

---

[199] A. Einstein, "Meine Antwort über die antirelativitätstheoretische G.m.b.H." (Gesellschaft mit beschrankter Haftung), *Berliner Tageblatt und Handelszeitung*, August 28, 1920, Nr. 402, pp. 1-2.

[200] *Ibid.*

[201] *Ibid.*

[202] *Ibid.*

The Einstein's article did not defuse the campaign against him and his theory: in fact, it even made the movement much stronger. Lenard felt deeply offended. Both parties, hostile and favourable to Einstein, announced that a discussion of the scientific value of the theory of relativity would be held at the 68th Congress of German Naturalists in Bad Nauheim.

## 3.4  Preparations for an extensive presentation of the new ether concept

When the anti-Einstein campaign started, Einstein, encouraged by Lorentz, was finishing work on his extensive presentation of the new conception of the ether. The correspondence preserved to this day proves that Lorentz tried to convince Einstein to complete the work, both in his letters, and during Einstein's visit to Leiden in October 1919. But let us return to 1916. When he received a letter from Einstein, in which the equality sign between the space-time continuum, the state of which was described by the metric tensor $g_{\mu\nu}$, and the ether appeared, Lorentz did not give in to the argument for rejecting the stationary ether, and remained faithful to his ether until his death in 1928. We cannot say, however, that he made no attempt to change his concept of the ether under the influence of his discussions with Einstein. After 1916, Lorentz tried from time to time to convince Einstein to deal with the ether in some more extensive way, and Einstein wrote to him on November 15, 1919:

> I will explain my position concerning the question of the ether in detail as soon as I find the opportunity to do so.[203] **d13**

The opportunity occurred when Lorentz offered Einstein a position as commuting professor to give lectures at the University of Leiden. Einstein accepted Lorentz's offer and decided to devote his inaugural lecture to the problem of the ether. On January 12, 1920 he wrote to Lorentz and Ehrenfest, who were both in Leiden at that time. The following excerpt is from the letter to Lorentz:

> I will give the inaugural lecture, which you mentioned, on the ether. It is a good opportunity to undertake the clarification that you suggested.[204] **d14**

In the letter to Paul Ehrenfest he wrote the following:

---

[203] A. Einstein, "Letter to H. A. Lorentz, 15/11/1919," *EA*, 16-494.
[204] A. Einstein, "Letter to H. A. Lorentz, 12/1/1920," *EA*, 16-498.

I will hold my inaugural lecture on "ether and the theory of relativity," because during my visit to Leiden Lorentz expressed the wish that I should take a position in public on this subject.[205] **d15**

Two months later, on March 18, 1920 Einstein wrote to Lorentz:

I am busy writing my inaugural lecture about the ether, which naturally cannot be anything different from a more or less coloured retrospective on the development of our conceptions of the physical properties of space. I hope that we have no substantially different opinions about these fundamental things.[206] **d16**

Einstein's inaugural lecture was to have been delivered on May 5, 1920, and Springer Verlag publishing house in Berlin printed it with that date of delivery. In reality, the lecture was delivered on October 17, 1920, after a discussion had taken place between Einstein and Lenard in Bad Nauheim.[207]

## 3.5     The Einstein–Lenard debate in Bad Nauheim

From September 19 to 25, 1920 the anticipated Congress of German Naturalists was held in Bad Nauheim. A number of papers were devoted to the relativity theory and Einstein's, Weyl's, and Mie's new theories of gravitation. Weyl presented his attempt to generalise the General Theory of Relativity. In his lecture, titled "Electricity and Gravitation,"[208] he referred to his version of the relativistic ether, pointing to the fact that the metrics of space-time continuum which represented the state of the ether unambiguously defined the field of gravitation:

The metric ("the state of the ether bearing the character of a field") unambiguously defines the affine connection ("gravitational field").[209] **d17**

M. von Laue, Gustav Mie, Leonhard Grebe, Hugo Dingler, and others delivered lectures as well. Grebe[210] presented two types of experiments which confirmed the shift in the solar radiation spectrum line towards the red. This was one of the effects predicted by Einstein's

[205] A. Einstein, "Letter to P. Ehrenfest, 12/1/1920," *EA*, 9-463.

[206] A. Einstein, "Letter to H. A. Lorentz, 18/3/1920," *EA*, 16-506.

[207] A. Pais, *op. cit.*, p. 313.

[208] H. Weyl, "Elektrizität und Gravitation," *PhZ*, 21 (1920), pp. 649-651.

[209] *Ibid.*, p. 649.

[210] L. Grebe, "Über die Gravitationsverschiebung der Fraunhoferschen Linien," *PhZ*, 21 (1920), pp. 662-666.

General Theory of Relativity. Dingler,[211] on the other hand, tried to challenge the basic principles and theorems of the Special Theory of Relativity and the General Theory of Relativity with the use of Lenard's *Relativity Principle, Ether, Gravitation* which he called "very thorough and deep" (*tiefschürfende Schrift*).

During the congress, there was also a general debate, mainly between Lenard and Einstein, which was later published in *Physikalische Zeitschrift*.[212] Lenard was the first to take the floor: after expressing his joy that the theory of gravitation presented by Weyl again included the ether, he repeated his chief accusations from 1917 targeting the generalised relativity principle. Einstein in his reply presented a new definition of the ether using the principle attacked by Lenard. He noted that physical space understood as the new ether was so constituted that no reference system or class of reference systems could be privileged, and therefore it was impossible to make a distinction between a "real" field of gravitation caused by masses and an "unreal" field caused by accelerated motion or rotation. Lenard was not convinced, however, as he rejected "unreal" fields, and pointed to a limitation on the relativity principle caused by rotation, which would lead—if the principle were accepted—to velocities exceeding the speed of light. We quote an excerpt from the debate:

> **Lenard**: In my article "On the principle of relativity, ether and gravitation" I expressed the opinion that the ether in some respects has not shown up, because it has not been treated in the right way. The principle of relativity works in a non-Euclidean space, which assumes different properties from place to place and with passing time; therefore precisely in space there must be something whose states generate these different properties and this something is the ether. I recognise the usefulness of the principle of relativity only when applied to the gravitational forces. For forces which are not proportional to the masses I consider it as not valid.

> **Einstein**: It is in the nature of the thing, that one can speak of the validity of the principle of relativity only if it holds with respect to all natural laws.

> **Lenard**: Only if one adds on appropriate fields. I mean that the principle of relativity can make new predictions only about gravitation, while the gravitational fields introduced in the case of forces which are not proportional to the masses do not introduce any

---

[211] H. Dingler, "Kritische Bemerkungen zu den Grundlagen der Relativitätstheorie," *PhZ*, 21 (1920), pp. 668-675.

[212] "Allgemeine Diskussion über Relativitätstheorie," *PhZ*, 21 (1920), pp. 666-668.

new point of view, other than to make the principle appear to be valid. I can add that the equivalence of all reference systems generates difficulties for the principle.

**Einstein**: In principle there is no system of coordinates which is privileged because of its simplicity; therefore there is also no method to discriminate between "real" and "unreal" gravitational fields. My second question is: What does the principle of relativity say about the forbidden thought experiment, in which for example the Earth is kept motionless while the rest of the universe is made to rotate around the earth's axis, so that superluminal velocities arise?

The first statement is not an affirmation, but a new definition for the concept of the "ether."

A thought experiment is one that can be performed at least in principle, even if not factually. It is useful for summarising real experiences in a clear way, so as to draw from them some theoretical consequences. A thought experiment is prohibited, only when it is in principle impossible to perform.

**Lenard**: I believe that I can summarise things thus: **1.** That it is better to avoid proclaiming the "abolition of the ether." **2.** That I still consider the reduction of the principle of relativity to a gravitational principle as demonstrated, and **3.**, that superluminal velocities seem really to create a difficulty for the principle of relativity; given that they arise in relation to an arbitrary body, as soon as they are attributed not to the body, but to the whole world, something which the principle of relativity in its simplest and heretofore existing form allows as equivalent.[213] **d18**

It can be seen from the above discussion that according to Einstein the sentence *"In principle there is no system of coordinates which is privileged because of its simplicity"* constituted *"a new definition for the concept of the ether."* This meant that Einstein had defined the new ether using the generalised principle of relativity. The sentence quoted above was not a justification, but a verbal expression of the generalised relativity principle.

It was also a new definition of the ether, because—according to Einstein—only space with no privileged reference system or privileged class of reference systems could be considered a new ether in his General Theory of Relativity.

---

[213] *Ibid.*, pp. 666-667.

## 3.6    Lenard's reaction to Einstein's response

Einstein's response to Lenard's objections did not satisfy the latter at
all, as can be seen from his speech summing up the discussion quoted
above, as well as in the "Appendix on the Nauheim Debate over the
Relativity Principle"[214] published in the revised edition of his work
*Relativity Principle, Ether, Gravitation*. A few remarks about the Appendix
are in order.

According to Lenard, the revised edition of his work contained some
ideas that were still valid. It was still up to date because in Nauheim
Einstein did not satisfactorily respond to the objections the work
contained. According to Lenard, Einstein started the debate with him
totally unprepared, although he must have known of the accusations, as
they had been published two years before. A theoretician who cannot
satisfactorily answer simple questions proves that his theory cannot be
recognised as free of faults.

From the Appendix, we also learn that:

> The "abolition of the ether" was again announced in Nauheim as a
> result, at a large open meeting, as it had been done previously by
> Einstein in Salzburg. (about the earlier announcement by Mr. Einstein
> himself at Salzburg, see the quotation in note 17 of p. 27). And
> nobody laughed. I do not know whether it would have been different
> if the abolition of the air had been announced.[215] **d19**

These last words underscore Lenard's unshaken belief in the
existence of the ether. For him, denying its existence was almost
equivalent to denying the existence of the air.

## 3.7    Weyl replies to Lenard's objections

A detailed report on the Nauheim conference was also prepared by
Hermann Weyl.[216] From this report we learn that the German
Mathematical Society did agree on the idea that the theory of relativity
was dealt with in numerous lectures at the conference, even joint
meetings of mathematicians and physicists. The first meeting and the
discussion between Lenard and Einstein, which Weyl would later call
"final and most dramatic," was chaired by Max Planck. Thanks to his

---

[214] Ph. Lenard, "Zusatz betreffend die Nauheimer Diskussion über das
Relativitätsprinzip," in: Ph. Lenard, *Über Relativitätsprinzip, Äther, Gravitation* (1921
edition), pp. 36-44.

[215] *Ibid.*, p. 37, footnote 1.

[216] H. Weyl, "Die Relativitätstheorie auf der Naturforscherversammlung in Bad
Nauheim," *Jahresbericht der Deutschen Mathematikervereinigung*, 31 (1922), pp. 51-63.

wonderful chairmanship skills and his impartiality, Planck was able to chair the meeting and discussion in such a way that the protagonists did not venture beyond a strictly scientific debate.

Weyl, who—it will be recalled—was a supporter of the relativistic ether, used the term "metric structure of ether" on a number of occasions in discussing his own lecture. Moreover, he also used the expression at a very special juncture, *i.e.*, in presenting Einstein's views and opinions. The following is typical:

> According to Einstein, the metric structure of ether is of the same kind as in Riemann's theory [...][217] **d20**

A significant part of Weyl's report dealt with the general discussion of the theory of relativity, which, as we know, was used mainly by Lenard to present his objections to Einstein. Here we quote a passage which also served as Weyl's answer to Lenard's accusations. In Weyl's view, the discussion between Lenard and Einstein focused on two issues:

> First issue: *the existence of the ether.* Lenard believes that in formulating the special theory of relativity, Einstein was precipitous in announcing the abolition of the ether. In fact, as he points out, today Einstein is once again speaking of an ether within the general theory of relativity. However the word should not mislead us about the difference in the thing! The old ether of the theory of light was a *substantial* medium, a three dimensional continuum, every point $P$ of which was at every moment $t$ in a well defined point of space $p$ (or in a well defined place in the universe); the recognisability of the same point of the ether at different times was the essential thing. Because of this ether, the four dimensional world becomes an infinite three dimensional continuum of one dimensional world-lines; for this reason it allows us to distinguish *rest* from *movement* in an absolute way. *In this sense*— Einstein stated nothing different—the ether is abolished by the special theory of relativity; it has been replaced by the affine geometrical structure of the world, which does not recognise a difference between rest and movement, but describes *uniform translation* as privileged in comparison with all other movements. The substantial ether was imagined by its inventors as something real and comparable to ponderable bodies. In the electrodynamics of Lorentz it was transformed into a purely geometrical structure, that is to say, rigid once and forever, and not influenced by matter. In Einstein's special theory of relativity it is replaced by a different structure, that of affine geometry. In the general theory of relativity, finally, the latter is transformed back into a "more affine connection" or "guiding field,"

---

[217] *Ibid.*, p. 52.

becoming a field of states endowed with physical reality and interacting with matter. It was for this reason that Einstein believed it appropriate to re-introduce the old word ether for a completely transformed concept; whether this was an appropriate choice, or not, is a question more of philological than physical interest.

Second issue: *superluminal velocity*. Lenard believes that the general theory of relativity reintroduces superluminal velocities, because it allows the use of the rotating Earth as reference system; in this way at sufficiently large distances superluminal velocities arise. This is an obvious misunderstanding. Let $x_1$, $x_2$, $x_3$ be space coordinates measured relative to the rotating Earth, and let $x_0$ be the relative "time" (its exact definition is not yet important). The coordinate lines $x_0$, on which for constant values of $x_1$, $x_2$, $x_3$, only $x_0$ is variable, do not all have a timelike direction, that is to say, with these coordinates one will not have in general $g_{00} > 0$. Still Einstein states that these systems of coordinates are also allowed; in addition, in such systems of coordinates his generally invariant gravitational laws are valid. Concerning this he holds firmly that *the universe line of a material body* always has a timelike direction, and that for a material body (and for a "signal velocity") no superluminal velocity can arise. Because of this a system of coordinates of this kind cannot be represented in its full extension by means of a "mollusc of reference";[218] that is to say, one cannot imagine any material medium, whose single elements describe the coordinate lines $x_0$ of that system of coordinates as universe lines.[219] **d21**

A few historical details about the Conference, reported by Weyl and Philipp Frank, need to be added. Weyl's report says that the last lecture was given by Rudolf. According to Weyl, Rudolf's lecture[220] represented a satirical or comical moment in the rather tragic course of the conference:

> So that the tragedy would not be without a satiric finale Mr. Rudolf developed a fantastic theory of the ether, with "cracks" between moving ether walls, with threads of stars, *etc.*, by means of which,

---

[218] The term "mollusk of reference" was used by Einstein himself *e.g.*, in his book: A. Einstein, *Über die Spezielle und Allgemeine Relativitätstheorie, Gemeinverständlich* (Braunschweig: F. Vieweg, 1917).

[219] H. Weyl, "Die Relativitätstheorie...," pp. 59-60.

[220] *Ibid.*, p. 63.

starting from nothing, he defined the mass of the sun precisely to an arbitrary number of decimals...[221] **d22**

Frank[222] reports that the Lenard-Einstein discussion, announced before the conference, attracted a few thousand people to Bad Nauheim. A sensational debate was anticipated. The building in which the conference was held was guarded by numerous policemen. Owing to Planck, who made every effort to ensure that the discussions immediately following the lecture would be complete and exhaustive, there was not much time left for the general discussion. Planck also made sure that the meetings finished as scheduled. When the clock struck 1 p.m., he interrupted the Lenard-Einstein debate with the following joke:

> Because the theory of relativity unfortunately has not yet succeeded in extending the absolute stationary 9 a.m. to 1 p.m, time of the meeting, the assembly must be adjourned.[223] **d23**

After the conference, the attacks on Einstein and his theory were no less frequent or less fierce. The attacks were led by Lenard, now a member of the national socialist party; this Nazi fanatic could not forgive Einstein's Jewish origin, his pacifism and the extraordinary popularity of his "absurd" theory, which contradicted the common sense of an experimental scientist.

## 3.8   Einstein's inaugural lecture in Leiden

Based on what Einstein wrote in his letters to Lorentz and Ehrenfest, his current views and opinions concerning the ether and, first of all, his new conception of the ether were to be covered in his lecture at Leiden.

In the first part of his lecture, Einstein briefly presented what, historically speaking, had caused physicists to introduce the concept of the ether. According to Einstein, there were two reasons for this: the problem of action at a distance and the discovery of the wave properties of light.

The problem of action at a distance appeared in connection with Newton's theory of gravitation. Prephysical thinking did not know objects that interact from a distance. In that mode of thought, all interactions were understood to be transmitted by direct contact. Falling of objects was treated as something very natural and was not associated

---

[221] *Ibid.*

[222] Ph. Frank, *op. cit.*, p. 275.

[223] *Ibid.*

with any action at a distance. It was only Newton's theory of gravitation that raised the problem of action at a distance. The dualism of direct reactions and distant interactions became unbearable for the human mind, which always seeks to learn the truth. Human cognition always strives for unity. There were two ways to eliminate this dualism:

1. Treat all reactions as distant interactions; the reactions thought of as contact reactions so far, would be treated as actions at a distance occurring when the distance is small.

2. Or treat Newton's reactions as apparent, assuming that the space in between contains a certain medium which transmits these reactions through its deformations. The second solution resulted in the hypothesis of the ether. Since the models of the ether that had been created were unsatisfactory and did not bring any progress in physics, people got used to treating Newton's law of universal gravitation as an irreducible axiom. This does not mean, however, that the ether hypothesis ceased to be of any interest to physicists. It retained its role instead, though often in an implicit form:

... the ether hypothesis was always bound to play a part in the thinking of physicists, even if it was mostly a latent hypothesis at first.[224] **d24**

The wave properties of light, discovered in the nineteenth century, were the second source of the ether hypothesis and contributed to its popularity. People started to think that light was a vibrational process of an elastic medium filling the whole world. The polarisability of light made people think that the medium should be thought of as a solid body, since transverse waves propagated in solid media, not in liquid ones. That is how the hypothesis of the "quasi-rigid" ether was born, an ether whose parts do not undergo any change, except for some minor strains, namely light waves. The theory was also called the theory of a resting ether; it gained some support in Fizeau's experiment, which led to the conclusion that the luminiferous ether does not take part in the motion of bodies. The phenomenon of light aberration was also an argument in support of the theory of a quasi-rigid ether.

The development of the theory of electromagnetism by Maxwell and Lorentz also contributed too, causing an unexpected change in our way of thinking about the ether. For Maxwell, the ether was just another thing with purely mechanical properties, albeit more complex than the mechanical properties of solid bodies. Although Maxwell, using his ether

---

[224] A. Einstein, *Äther und Relativitätstheorie* (Berlin: J. Springer, 1920), p. 4.

model, managed to discover the laws expressed by equations now known as the Maxwell equations, their mechanical interpretation left much to be desired. The laws were clear and simple but their mechanical presentation was difficult to understand and full of inconsistencies. It made physicists assume that, apart from the existence of purely mechanical phenomena, there also existed electromagnetic phenomena that were independent and that could be expressed by means of the field concept. Thus, a new intolerable dualism involving the essentials of physics was born. There were many attempts to eliminate the dualism, *e.g.*, Hertz's theory. None of them was successful, however.

Lorentz was the first to succeed in formulating a theory that was consistent with experience and which simplified the essentials to a great extent. Lorentz made wonderful progress in depriving the ether of mechanical properties, and at the same time depriving matter of electromagnetic properties. Both in empty space and inside material bodies, the ether was the only support for electromagnetic fields. Elementary particles of matter, which, according to Lorentz were the only particles that move, fulfilled their electromagnetic function only due to the fact that they were carriers of electric charge. He managed to express all electromagnetic phenomena by means of the Maxwell equations for the vacuum. In this new approach, the ether lost all of its mechanical properties except one:

> As to the mechanical nature of the Lorentzian ether, it may be said of this, in a somewhat playful spirit, that immobility is the only mechanical property of which it has not been deprived by H.A. Lorentz. It may be added that the whole change in the conception of the ether, which the special theory of relativity brought about, consisted of taking away from the ether its last mechanical quality, namely, its immobility.[225] **d25**

At this point the introduction to the lecture on the ether of Special Relativity ends.

The second part of Einstein's lecture began with the answer to the question: how did the Special Theory of Relativity deprive the ether of its last mechanical feature, namely its immobility? Einstein reiterated the arguments presented earlier in the *Morgan Manuscript*. The Maxwell and Lorentz theory of the electromagnetic field was used as a model for the Special Theory of Relativity, which represents a new approach. Let $K$ be a co-ordinate system in relation to which Lorentz's ether is at rest. The

---

[225] *Ibid.*, p. 7.

Maxwell-Lorentz equations are thus valid in this system. According to the Special Theory of Relativity, however, these equations were also valid in relation to the $K'$ system, in uniform, rectilinear motion in relation to $K$. This raises a delicate question: why, from among all systems that are physically equivalent, should we single out the one in relation to which the ether is at rest? This points to a theoretical asymmetry that is not matched by any experimental asymmetry. From the logical point of view, assuming the equivalence of the $K$ and $K'$ systems, the statement that the ether is at rest in relation to $K$ and in motion in relation to $K'$ can be viewed as possible. From the theoretical point of view, however, it is an unacceptable asymmetry, unjustified and unbearable for theoretical minds. To eliminate this asymmetry, the Special Theory of Relativity relied on the special principle of relativity. Under the circumstances, the assumption that the ether does not exist at all, and that electromagnetic fields are autonomous entities, seemed to be the best solution.

> More careful reflection teaches us, however, that denial of the existence of the ether is not demanded by the special principle of relativity. We may assume the existence of an ether; only we must give up ascribing a definite state of motion to it, *i.e.*, we must by abstraction take away from it the last mechanical characteristic that Lorentz had still left it.[226] **d26**

The ether of the Special Theory of Relativity was a spatial continuum, not, however, an entity consisting of points or particles whose behaviour could be followed in time. Every physical particle has its world-line in a four-dimensional universe. The ether of the Special Theory of Relativity, as an entity not composed of points or particles, appeared in the four dimensional approach as not consisting of world-lines:

> One can imagine that extended physical objects exist to which the idea of motion cannot be applied. They are not to be conceived as composed of particles, whose course can be followed separately through time. In Minkowski's language, this is expressed as follows: Not every extended entity in the four-dimensional world can be regarded as composed of world-lines. The special principle of relativity forbids us to assume that the ether consists of particles observable through time, but the ether hypothesis in itself is not in conflict with the special theory of relativity. However, we must take care not to ascribe a state of motion to the ether.[227] **d27**

---

[226] *Ibid.*, p. 9.
[227] *Ibid.*, p. 10.

In spite of this, from the point of view of the Special Theory of Relativity, the ether hypothesis seemed to be devoid of physical meaning. Electromagnetic fields were presented as autonomous entities, as formations not expressible in any other way. For this reason it seemed redundant to postulate the existence of a homogeneous isotropic ether whose states would be these fields.

> On the other hand there is a weighty argument to be adduced in favour of the ether hypothesis. To deny the existence of the ether means, in the last analysis, denying all physical properties to empty space.[228] **d28**

We cannot agree with the statement that empty space has no physical properties, because the facts known from the field of mechanics contradict it. Real properties of space had already been observed by Newton, and now the General Theory of Relativity presented these properties in a more complete form.

In this part of the lecture, Einstein briefly showed how the physical properties of empty space were discovered, which actually meant that space could not be treated as really empty. It turned out that the behaviour of a freely moving system of bodies in empty space depended on the relative position (distance) and relative velocity of its elements as well as its rotation state. In order to treat rotation of a system as something real, at least from the formal point of view, Newton made space objective. Since he considered absolute space to be a real entity, he could also consider the rotation of a system in relation to that absolute space to be something real. According to Einstein, Newton might just as well have called his absolute space the ether, because it possesses real properties that determine the behaviour of bodies. It is a significant feature of Newton's theory that apart from observable objects, one had to assume the real existence of something which, although not observable, was nevertheless necessary if one wished to consider acceleration and rotation as real.

To avoid the need to treat something that was not observable as something real, Mach explained acceleration and rotation as taking place not in relation to absolute space, but as an average acceleration and rotation in relation to the totality of distant masses of the rest of the world. But inertial resistance, contradicting acceleration in relation to distant masses, assumed unobservable distant interactions without any medium at all. Since a modern physicist could not accept such

---

[228] *Ibid.*, p. 11.

interactions, he would have to refer back to the ether as a medium for transmitting interactions that create inertial resistance and cause the appearance of forces in rotating bodies and systems of bodies. As we can see, Einstein at that time clearly saw the need for a medium to transmit interactions from distant masses. This medium, however, was essentially different from the previously assumed ethers.

> This conception of the ether to which Mach's approach leads, differs in an important respect from that of Newton, Fresnel and Lorentz. Mach's ether not only conditions the behaviour of inert masses but is also conditioned, as regards its state, by them.[229] **d29**

The fourth part of Einstein's lecture shows how Mach's ideas, outlined above, were implemented and developed in the General Theory of Relativity.

> Mach's ideas find their full development in the ether of the general theory of relativity. According to this theory the metrical properties of the spacetime continuum in the neighborhood of separate spacetime points are different and simultaneously conditioned by matter existing outside the region in question. The spatio-temporal variability of relations of rods and clocks to one another, or the knowledge that "empty space" is, physically speaking, neither homogeneous nor isotropic—which compels us to describe its state by means of ten functions, the gravitational potentials $g_{\mu\nu}$, has no doubt finally disposed of the notion that space is physically empty. But this has also once again given the ether notion a definite content—though one very different from that of the ether of the mechanical wave-theory of light. The ether of the general theory of relativity is a medium which is itself devoid of all mechanical and kinematic properties, but it helps to determine mechanical (and electromagnetic) phenomena.[230] **d30**

Here Einstein presented what was new about the General Theory of Relativity ether, as compared to the Lorentzian ether. The difference between the two theories was that the state of the ether in the General Theory of Relativity at each point was defined by theoretically available real relationships with matter and with the state of the ether at adjacent points. The state of Lorentz's ether, on the other hand, in the absence of electromagnetic fields, could only be defined by itself and was the same everywhere. We might say, then, that the General Theory of Relativity ether emerged from Lorentz's ether when the latter was made relative.

---

[229] *Ibid.*, p. 12.
[230] *Ibid.*

For Einstein, there was a clear difference between the relationship {general relativistic ether – gravitational field} and the relationship {general relativistic ether – electromagnetic field}. As for the first relation:

> No space and no portion of space [can be conceived of] without gravitational potentials; for these give it its metrical properties without which it is not thinkable at all. The existence of the gravitational field is directly bound up with the existence of space.[231] **d31**

As we can see, physical space, insofar as it possesses metrical structure, is here identified with the gravitational field. Because it has real metrical properties which, according to the General Theory of Relativity, are physical properties (gravitational potentials), it constitutes the ether of the General Theory of Relativity, which Einstein called the gravitational ether. The relationship between the ether of the General Theory of Relativity and the electromagnetic field is totally different:

> On the other hand a portion of space without an electromagnetic field is perfectly conceivable; hence the electromagnetic field, in contrast to the gravitational field, seems in a sense to be connected with the ether only in a secondary way, inasmuch as the formal nature of the electromagnetic field is by no means determined by the gravitational ether. In the present state of theory it appears that the electromagnetic field, as compared with gravitational field, was based on a completely new formal motive; as if nature, instead of endowing the gravitational ether with fields of the electromagnetic type, might equally well have endowed it with fields of a quite different type, for example, fields with a scalar potential.

> Since according to our present-day notions the primary particles of matter are also, fundamentally, nothing but condensations of the electromagnetic field, our modern representation of the world recognises two realities which are conceptually quite independent of each other even though they may be causally connected, namely the gravitational ether and the electromagnetic field, or—as they might be called—space and matter. It would, of course, be a great step forward if we succeeded in combining the gravitational field and the electromagnetic field into a single homogeneous structure.[232] **d32**

Einstein would briefly mention the efforts made by Weyl:

---

[231] *Ibid.*, pp. 13-14.
[232] *Ibid.*, p. 14.

An exceedingly brilliant attempt in this direction has been made by the mathematician Hermann Weyl; but I do not think that it will stand up to the test of reality.[233] **d33**

The lecture summary actually refers only to the part of the lecture which dealt with the ether of the General Theory of Relativity. The text is as follows:

> We may sum up as follows: According to the general theory of relativity, space is endowed with physical qualities; in this sense, therefore, there exists an ether. According to the general theory of relativity, space without ether is unthinkable; for in such space, not only would there be no propagation of light, but also no possibility of existence for standards of space and time (measuring rods and clocks), nor therefore any space-time intervals in the physical sense. But, this ether may not be thought of as endowed with the quality characteristic of ponderable media, to consist of parts that may be tracked through time. The idea of motion may not be applied to it.[234] **d34**

This is how Einstein concluded his first major work on the new ether. We need only add that the remark to the effect that the new ether did not consist of parts observable in time was not only an indication of one of the features of the new ether, but also a declaration against Lenard's concept of ether. Lenard often emphasised in his work[235] that his ether consisted of parts. Einstein definitely rejected the existence of this kind of ether.

## 3.9    Eddington's relativistic ether

Arthur Eddington was captivated by the idea of the relativistic ether from the very beginning. In 1920, he published a book popularising the General Theory of Relativity,[236] in which, among other things, he presented and promoted the new ether in his own way. Having emphasised that in the theory of relativity the point event is a key concept and that a set of all such events was called the world, he wrote:

> What we have here called the *world* might perhaps have been legitimately called the *aether*; at least it is the universal substratum of

---

[233] *Ibid.*, pp. 14-15.

[234] *Ibid.*, p. 15.

[235] Ph. Lenard, *Über Äther und Materie*, 2 Aufl. (Heidelberg: Carl Winters Universitätsbuchhandlung, 1911).

[236] A.S. Eddington, *Space, Time and Gravitation. An Outline of the General Relativity Theory* (Cambridge: Cambrige Univ. Press, 1920).

things which the relativity theory gives us in place of the [old—L.K.] aether. [...] matter is some state in the aether [...], matter (or electromagnetic energy) is the only thing that can have a velocity relative to the frame of reference. The velocity of the world-structure or aether, where the $T_{\mu\nu}$ [matter tensor—L.K.] vanish, is always of the indeterminate form $0 \div 0$. On the other hand acceleration and rotation are defined by means of the $g_{\mu\nu}$ and exist wherever these exist; so that the acceleration and rotation of the world-structure or aether relative to the frame of reference are determinate.[237] **d35**

While Einstein liked Eddington's book, he also made some critical remarks about it. In his letter to Sommerfeld, Einstein wrote:

Eddington's book is really very brilliant. But I cannot accept the tendency to consider the laws of nature only as schemes of order for the subdivision of cases that fall in them, or that do not fall. It is also necessary to criticise the fact that he often describes the theory of relativity as *logically* necessary. God could also have decided to create an absolute static ether instead of the relativistic ether. This would hold especially, if he were to adapt the ether to the (substantial) independence from matter, as in de Sitter, an opinion toward which Eddington obviously leans; because in such a case an "absolute" function should also be attributed to the ether. It is remarkable that in most heads there is no organ to assess this state of things.[238] **d36**

In 1921 Eddington developed the relativistic concept of the ether in his own way, making a second attempt, after Weyl's, to unify gravitational and electromagnetic interactions. His approach relied on affine geometry. In this geometry, connection, and not metric, is considered to be the basic mathematical entity. The metric $g_{\mu\nu}(x)$ needed for the description of the gravitational interactions, appears here as something secondary, which is derived from connection.[239] In this new development of the theory of relativity, Eddington presented the relativistic ether as a field in which both interactions, gravitational and electromagnetic, took place.

---

[237] *Ibid.*, pp. 187, 194 -195.

[238] A. Einstein, "Letter to A. Sommerfeld, 28/11/1926," in: *A. Einstein, A. Sommerfeld Briefwechsel*, red. and comments by A. Hermann (Basel-Stuttgart: Schwabe u. Co. Verlag, 1968), p. 109.

[239] H. Arodź, "Albert Einstein a problem unifikacji oddziaływań fundamentalnych" (Albert Einstein and the problem of unification of fundamental interactions), *PF*, 37 (1986), p. 305.

[...] the field (or the ether) maintains two characteristic properties, the gravitational potential (or metrics) and the electromagnetic force.[240]
**d37**

Eddington's attempted unification, like Weyl's, was criticised by Einstein. In his opinion, in fact, it was incompatible with the experimental evidence.[241]

## 3.10  Weyl's improved version of the relativistic ether

In proposing his version of the relativistic ether, Weyl started to treat the relativistic ether as one field in which both gravitational and electromagnetic interactions took place. He abandoned the dualism of ether and field, discussed in his book *Raum, Zeit, Materie*, mentioned earlier. It is clearly stated in his work *Feld und Materie*,[242] published in 1921, where he wrote:

> This whole conception leads to an extraordinary unity, if the ether is no longer composed of two fields in an extended medium having no mutual relation, yet dominating the external world, but is itself this medium endowed with a matter-dependent metrical structure. This is the teaching of the theory of relativity extended by me, which I have here presented.[243] **d38**

## 3.11  Kaluza's pentadimensional world

In 1921, Theodor Kaluza published his attempt to unify both interactions.[244] A hypothesis that the space-time continuum was by its very nature not four-dimensional but five-dimensional was presented there for the first time. The geometry of this five-dimensional space-time continuum was identical with the geometry used in the General Theory of Relativity, the only difference being the number of dimensions. According to Kaluza, the metric tensor of his five-dimensional time-space continuum was a symmetric $5 \times 5$ matrix and took the following form:

---

[240] A. S. Eddington, *Relativitätstheorie in mathematischer Behandlung* (Berlin: Verlag v. J. Springer, 1925), p. 334.
[241] A. Einstein, "Über den Äther," *VSNG*, 105 (1924), p. 91.
[242] H. Weyl, "Feld und Materie," *AdP*, 65 (1921), pp. 541-563.
[243] *Ibid.*, p. 559.
[244] Th. Kaluza, *SPAW* (1921), p. 966; note taken from the article by H. Arodź, *op. cit.*, p. 309.

$$g_{AB}(X) = \left( \frac{g_{\mu\nu}(X)A_{\mu}(X)}{A_{\mu}(X)g_{44}(X)} \right)$$

where $\mu, \nu = 0, 1, 2, 3$ and $A, B = 0, 1, 2, 3, 4$; $g_{\mu\nu}$ describes the gravitational field, and $A_{\mu}$ the electromagnetic field.[245]

Einstein recognised the importance of the problem of unification. Starting from Weyl's first attempt, he had been treating it as one of the most important problems of theoretical physics. In spite of this, however—or maybe because of it—he did not like any of the attempts at unification that had been made, and that is why, in 1923, he started publishing works criticising them. Together with Jakob Grommer, he published a work in which he indicated the deficiencies of Kaluza's model. According to the two authors, it was mainly the principle of covariance of physical properties and invariance of field equations in relation to general changes of co-ordinates in five-dimensional space-time continuum that did not have any physical justification.[246]

In 1923, Einstein also published four works[247] criticising Eddington's attempts at unification. According to Einstein, Eddington's model was non-physical because, among other things, it was not able to account for the difference between the masses of "positive and negative electrons."[248] At that time, "positive electrons" meant protons.

## 3.12   Einstein's second major work on the new ether

Einstein's essay "On the Ether,"[249] published in Switzerland, is his second extensive work on new ether. The essay first of all emphasised the ether's active role in physical processes. It completed the first stage of the development of the new concept of ether. The subsequent stages will be discussed in the next chapter.

---

[245] H. Arodź, *op. cit.*, p. 304.

[246] *Ibid.*, p. 306.

[247] *Ibid.*, p. 305-306.

[248] *Ibid.*, p. 306.

[249] A. Einstein, "Über den Äther," *VSNG*, 105 (1924), pp. 85-93, see p.85. English translation: "On the Ether," in: *The Philosophy of Vacuum*, edited by S. Saunders and H. R. Brown (Oxford: Clarendon Press, 1991), pp. 13-20.

# Über den Äther

Von

Prof. Dr. Albert Einstein (Berlin)

Wenn hier vom Äther die Rede ist, so soll es sich natürlich
nicht um den körperlichen Äther der mechanischen Undulationstheorie handeln, welcher den Gesetzen der Newtonschen Mechanik
unterliegt, und dessen einzelnen Punkten eine Geschwindigkeit
zugeteilt wird. Dies theoretische Gebilde hat nach meiner Überzeugung seit der Aufstellung der speziellen Relativitätstheorie seine
Rolle endgültig zu Ende gespielt. Es handelt sich vielmehr allgemeiner um diejenigen als physikalisch-real gedachten Dinge, welche
neben der aus elektrischen Elementarteilchen bestehenden ponderabeln
Materie im Kausal-Nexus der Physik eine Rolle spielen. Man könnte
statt von „Äther" also ebensogut von „physikalischen Qualitäten des
Raumes" sprechen. Nun könnte allerdings die Meinung vertreten
werden, dass unter diesen Begriff alle Gegenstände der Physik
fallen, weil nach der konsequenten Feldtheorie auch die ponderable
Materie, bzw. die sie konstituierenden Elementarteilchen als „Felder"
besonderer Art, bzw. als besondere „Raum-Zustände" aufzufassen
seien. Indessen wird man zugeben müssen, dass beim heutigen
Stande der Physik eine solche Auffassung verfrüht wäre, denn bisher sind alle auf dies Ziel gerichteten Bemühungen der theoretischen
Physiker gescheitert. So sind wir beim heutigen Stand der Dinge
faktisch gezwungen, zwischen „Materie" und „Feldern" zu unterscheiden, wenn wir auch hoffen dürfen, dass spätere Generationen
diese dualistische Auffassung überwinden und durch eine einheitliche ersetzen werden, wie es die Feldtheorie in unseren Tagen
vergeblich versucht hat.

Man glaubt gewöhnlich, dass die Physik Newtons keinen Äther
gekannt habe, sondern dass erst die Undulationstheorie des Lichtes

**Figure 3.** The first page of Einstein's paper *Über den Äther* published in 1924. This was
Einstein's second major work on the new ether. (Reproduced with the kind permission of
The Albert Einstein Archives, The Jewish National and University Library, The Hebrew
University of Jerusalem, Israel.)

To make absolutely clear what kind of ether was to be understood,
Einstein began the text with the following words:

If we are here going to talk about the ether, we are not, of course, talking about the physical or material ether of the mechanical theory of undulations, which is subject to the laws of Newtonian mechanics. This theoretical edifice has, I am convinced, finally played out its role since the setting up of the special theory of relativity. It is rather more generally a question of those kinds of things that are considered as physically real, which play a role in the causal nexus of physics, apart from the ponderable matter that consists of electrical elementary particles. Therefore, instead of speaking of an ether, one could equally well speak of physical qualities of space.[250] **d39**

In this text, Einstein emphasised the view that he had not stressed so much in his inauguration lecture in Leiden. In the new ether concept, we find an activation of space (which was introduced as early as 1907, *i.e.* with the first attempt to generalise the theory of relativity, and which was fully rendered in the final General Theory of Relativity). Thus, space was no longer treated as a passive container of matter and events. The above quoted text was a general definition of the new ether, in which its active role was emphasised. According to Einstein, the definition was so general that it also included Newton's "absolute space," whose active role in cause-effect relations had been disregarded by the majority of physicists. According to Einstein, Newton's "absolute space" actively determined the inertial behaviour of bodies and, because of its active role, it could not be ignored in Newton's definition of force ($F = ma$). In Einstein's opinion, Newton's real and active space should be called the ether of mechanics.

We are going to call this physical reality, which enters into Newton's law of motion alongside the observable ponderable bodies, the, "ether of mechanics."[251] **d40**

Having introduced the new concept of ether in 1916, Einstein admitted the reality of physical space. He had thus come a long way away from the epistemological views of Mach and other positivists who considered space (especially "*absolute space*") to be a metaphysical addition that should be removed from physics. It is also significant that in this later work, Einstein did not even mention "Mach's ether," which he had spoken about in his lecture in Leiden, and which he then thought to be an anticipation of the ether of his General Theory of Relativity. Now he preferred Newton's space, which he called the ether of mechanics, and

---

[250] *Ibid.*, p. 85.
[251] *Ibid.*

which he thought to be the first step towards his General Theory of Relativity active ether.

Why this turn of events? In 1921, after Mach's death, his book *Principles of Physical Optics*[252] was published. In the preface to that work, written in 1913, Mach strongly objected to being treated as a forerunner of relativistic theory, and was quite opposed to the theory because (as he said) of its increasing dogmatism.[253] Einstein realised that it was an exaggeration to treat the hypothetical existence of the ether, as Mach had presented it, as the model of the ether which had anticipated the ether of the General Theory of Relativity; similarly, he had stated that Mach had almost managed to formulate the General Theory of Relativity, which Einstein wrote about in his article dated 1916 to honour Mach's scientific achievement after his death. After quoting a passage from Mach's *Mechanics*, Einstein wrote:

> The quoted words show that Mach had clearly recognised the weaknesses of classical mechanics, and that he was not very far from formulating the general theory of relativity, already half a century ago![254] **d41**

The fact that Mach condemned the theory of relativity was a very unpleasant experience for Einstein. He stopped praising Mach's achievements and started criticising him and his epistemological views. To demonstrate this, we quote Einstein's answer to the question asked by Emil Meyerson during a reception on April 6, 1922 in Paris, organised by the French Philosophical Society to honour Einstein. Einstein replied as follows:

> There does not appear to be a close relation from the logical point of view between the theory of relativity and Mach's theory. For Mach, there are two points to distinguish: on the one hand there are the immediate data of experience, things we cannot touch; on the other, there are concepts which we can modify. Mach's system studies the existing relations between data of experience; for Mach science is the totality of these relations. That point of view is wrong, and, in fact, what Mach has done is to make a catalogue, not a system. To the extent that Mach was a good mechanician he was a deplorable philosopher.[255] **d42**

---

[252] E. Mach, *Die Prinzipien der Physikalischen Optik* (Leipzig: J.A. Barth, 1921).

[253] M. Heller, "Jak Einstein...," p. 19.

[254] A. Einstein, "Ernst Mach," *PhZ*, 17 (1916), p. 103.

[255] A. Einstein, a note in: *Nature*, 112 (1913), Nr. 2807, p. 253.

However, let us return to Einstein's work *On the Ether* of 1924. In one of its later sections, Einstein said that the activity of the ether of Newtonian mechanics was of a one-way nature, because the ether itself was not subject to any reaction and, in this sense, was absolute.

The Special Theory of Relativity was a step forward. Its ether, like Newton's "ether of mechanics," did not distinguish any definite state of motion. In the Special Theory of Relativity, the absolute character of time and simultaneity disappeared, and that was why the Special Theory of Relativity became four-dimensional. The ether was a real active space-time continuum that determined the inertial behaviour of bodies and the propagation of light. The Special Theory of Relativity ether also conditioned the geometry of bodies, which Newton's mechanical ether had not done. It was, however, like the latter, an absolute ether in the sense that it did not depend on the distribution and motion of matter.

> Also, following the special theory of relativity, the ether was absolute, because its influence on inertia and light propagation was thought to be independent of physical influences of any kind. [...] Therefore the geometry of bodies is influenced by the ether as well as the dynamics.[256] **d43**

The General Theory of Relativity was the next step forward in the development of the relativistic concept of the ether. The General Theory of Relativity ether determined not only the inertial behaviour of bodies, but also their gravitational behaviour. The action of the General Theory of Relativity ether was thus of an inertial-gravitational character. Hence, the General Theory of Relativity ether was no longer absolute, because its structure depended on the presence and motion of matter. Its metric properties changed from place to place, and also in time.

> The ether of the general theory of relativity therefore differs from that of classical mechanics or the special theory of relativity, in so far as it is not "absolute," but is determined in its locally variable properties by ponderable matter.[257] **d44**

The last step in the development of the relativistic concept of the ether would be the creation of a unified field theory in which a unification of gravitational and electromagnetic interactions is achieved and in which matter consisting of particles would constitute special states of physical space. Thus far, the attempts to develop such a theory have been unsuccessful, the reason lying not in physical reality, but in the

---

[256] A. Einstein, "Über den Äther," pp. 89-90.
[257] *Ibid.*, p. 90.

deficiencies of our theories. It would be ideal to develop such a unified field theory in which all the objects of physics would come under the concept of the ether. Einstein pointed out this problem at the very beginning of his article:

> [...] one can defend the view that this notion [the ether—L.K.] includes all objects of physics, since according to a consistent field theory, ponderable matter and the elementary particles from which it is built also have to be regarded as "fields" of a particular kind or as particular "states" of space.[258] **d45**

Despite the unsuccessful attempts to develop a unified field theory, one thing was sure—space has real physical properties that take an active part in physical processes—and this is why no complete physical theory could fail to include something that Einstein called "influence of the environment" (*Milieu-Einflüsse*). In other words, no complete physical theory could function without the ether, although there were some branches of physics that were not complete physical theories and that operated without the ether, namely: Euclidean geometry understood as a branch of physics, non-Euclidean geometries of constant curvature, and classical kinematics.

At the end of his article, Einstein mentioned the latest achievements in the developing theory of quanta. He expressed the view that even though it was likely to be developed into a mature theory of quanta, theoretical physics would never be able to function without the ether, understood as a space-time continuum possessing physical properties:

> But even if these possibilities should mature into genuine theories, we will not be able to do without the ether in theoretical physics, *i.e.* a continuum which is equipped with physical properties; for the general theory of relativity, whose basic points of view surely will always maintain, excludes direct distant action. But every contiguous action theory presumes continuous fields, and therefore also the existence of an "ether."[259] **d46**

## 3.13    Evolution of Einstein's epistemological views

As we know, the young Einstein was fascinated by the epistemological views of the positivists, and to a certain extent became their supporter. During the discussions with the friends (Solovine, Habicht) with whom he had established the Olympus Academy, he also

---

[258] *Ibid.*, p. 85.
[259] *Ibid.*, p. 93.

became acquainted with the works of the representatives of the first positivism (David Hume, John Stuart Mill) as well as the second positivism of Ernst Mach, Wilhelm Ostwald and Richard Avenarius. Later, however, reviewing his own scientific achievements and those of other distinguished physicists, he gradually started to abandon the positivistic views he had originally professed. He would, for example, change his opinions concerning the role of induction in physics.

The discussions held at the meetings of the Olympus Academy on the third book of John Stuart Mill's *Logic* concerning induction convinced Einstein of the essential character of induction and of its primacy over deduction in natural sciences. Later, however, influenced by considerations of the methods of theoretical physics, Einstein would change his views quite fundamentally, and would keep emphasising the primacy of deduction over induction in human cognitive processes as well, anticipating to a great extent the views of Karl Popper, the famous methodologist of physics.

In 1914, before the General Theory of Relativity was finally formulated, in his inauguration speech[260] upon his induction into the Prussian Academy of Sciences, Einstein presented deduction as the basic method of theoretical physics. According to Einstein, the work of a theoretical physicist consisted of two stages. During the first stage, through his own creativity, he discovered the basic postulates or principles, and, in the second stage, he built up a theory using deduction as the basic method. The theory thus developed was then experimentally tested. The essential role of deduction was similarly stressed in a speech[261] he made in 1918 to honour Planck's 60th birthday.

The views closest to Popper's later opinions were presented by Einstein in a very interesting but little-known article (not mentioned in any reference list of his works)[262] published in a Berlin newspaper, *Berliner Tageblatt*, on Christmas Day, 1919. The editors asked Einstein to express his political views on the difficult situation of Germany after the First World War. Rather than make political statements, Einstein sent a politically neutral article entitled "Induction and Deduction in Physics,"[263] which ended with the following words:

---

[260] A. Einstein, "Antrittsrede," *SPAW*, 2. Tl. (1914), pp. 739-742.

[261] A. Einstein, "Motiv des Forschers," in: *Zu Max Plancks 60. Geburstag: Ansprechen in der Deutschen Physikalischen Gesellschaft* (Karlsruhe: Müller, 1918), pp. 29-32.

[262] Prof. Don Howard informed me of the existence of this article.

[263] A. Einstein, "Induktion und Deduktion in der Physik," in: *Berliner Tageblatt und Handelszeitung*, morning edition, Nr 617, 25/12/1919, Suppl. 4.

I bring to the reader in this period of excitement this small, objective, dispassionate consideration, because I have the opinion, that through calm devotion to the eternal objectives, which are common to all civilised human beings, one can more efficiently serve the recovery of political health than by political considerations and declarations.[264] **d47**

Since the methodological remarks expressed in that article would significantly influence Einstein's works on the conceptions of space, ether and field discussed in the next chapter, let us dwell on them for a while. In the first part of the article, Einstein presented the view he had supported when still under the influence of John Stuart Mill:

> The simplest representation that one can imagine of the birth of an empirical science is in agreement with the inductive method. Individual facts are chosen and assembled, and a regular connection between them comes out clearly. Through the arrangement of these regularities one can again obtain more general regularities, until a more or less unitary system is elaborated for the available set of single facts. This must be of a such a kind that retrospective thinking can start from the generalisations so obtained and recover the individual facts with a reverse method of pure reasoning.[265] **d48**

According to Einstein, the origin of experimental sciences presented in the above quoted passage did not match reality. The most significant achievements in the natural sciences were made using the method opposite to that of induction.

> Already a superficial look at real developments teaches us that the great steps forward in scientific knowledge evolved only in small part like this. If the scientist, in fact, approached problems without some preconceived opinion, how could he extract from the huge number of facts of the most complex experience cases that are simple enough to let rational connections become evident? Galilei would never have been able to find the law of free fall without the preconceived opinion that the falls we observe in reality are complicated by effects due to the resistance of air, and that one must also imagine cases, in which it possibly plays a small role.
>
> The really great steps forward in the knowledge of nature were made in a way almost diametrically opposed to induction. The intuitive understanding of what is essential in a large complex of facts leads the scientist to the construction of one or several hypothetical fundamental laws. From the fundamental law (system of axioms) he

---

[264] *Ibid.*

[265] *Ibid.*

draws the consequences, as much as possible in a completely deductive way. These consequences, often deduced from the fundamental law by means of complicated developments and calculations, can be compared with experience and so provide a criterion for the acceptance of the fundamental law. Fundamental laws (axioms) and consequences form together what is called a "theory." Every expert knows that the greatest steps forward in the knowledge of nature, for example Newton's theory of gravitation, thermodynamics, kinetic theory of gases, modern electrodynamics, *etc.*, were all born in this way, and that to their foundations must be attributed that hypothetical character as a matter of principle. Therefore, the researcher certainly starts always from the facts, whose connection is the task of his work. He does not, however, reach his theoretical system in a methodical and inductive way, rather, he clings to the facts by intuitive selection among the thinkable theories based on the axioms.[266] **d49**

A notable "anti-inductionism" and "hypothetism" of Einstein's methodological views can be seen in the above quoted passage. These views anticipated Popper's "anti-inductionism" and "hypothetism." The remark that comparing conclusions drawn deductively from hypothetically assumed laws (axioms) to experience only justified the laws, but did not exclude other theories with other axioms, would be reflected in Popper's views. Einstein also anticipated Popper's view that each theory could always be considered false (theory falsification) but could never be proved to be true. The following passage justifies the statement:

> A theory can, in fact, be recognised as false if there is a logical error in its deductions, or as falsified, if a fact is not in agreement with one of its consequences. But the truth of a theory can never be proved, because one never knows if, in the future, some experience will become known which contradicts its deductions. Other systems of thought can always be imagined which are able to link the same existing facts. If there are two theories which are both compatible with the given empirical material, there is no other criterion for the preference of one or the other, than the intuitive picture of the scientist. In this way one can understand that sagacious scientists who dominate theories and facts, can be ardent followers of opposite theories.[267] **d50**

---

[266] *Ibid.*
[267] *Ibid.*

Note that the passages quoted above comprise the entire article by Einstein.

Today there is no doubt at all about the deductive character of the Special Theory of Relativity and the General Theory of Relativity. The experts on both theories emphasise it. For example, Landau and Lifshitz would write of the General Theory of Relativity: "… it was built up by Einstein in a purely deductive way and … it was not until later that it was confirmed by astronomical observations."[268]

Einstein realised the deductive character of both theories quite early, and this is why his "anti-inductionism" and "hypothetism" appeared in the first works in which he expressed his epistemological views. In the inaugural speech he gave upon receiving membership in the Prussian Academy of Sciences, Einstein stressed the deductive character of the Special Theory of Relativity. The theory was deduced from two assumptions, which—we might add—had been clearly presented in his work dated 1905: namely, from the special principle of relativity, according to which all inertial systems were considered equal in formulating laws of mechanics, electrodynamics and optics, and from the principle of constant velocity of light in these inertial systems. In the *Morgan Manuscript* written at the turn of 1918/19, he presented three further assumptions of the Special Theory of Relativity, unconsciously adopted at first, namely: uniformity of space and time, isotropy of space and independence of measurement bars and clocks of their history.[269] Einstein would consciously use the deductive method in his attempts to formulate a unified relativistic field theory—attempts which are closely tied to the development of his new ether.

Einstein's departure from the inductionism of the positivists took place quite early, but his separation from the second positivistic view, according to which the ether (Ostwald) and space (Mach) are only metaphysical concepts that should be removed from physics, did not occur until June of 1916. In this year Einstein became convinced that the space-time continuum of his General Theory of Relativity had objectively real properties: from this idea arose the new ether, which gradually gained strength in his mind. Einstein would emphasise it in his 1918 work, in which his new ether concept was published for the first time:

[268] L. D. Landau, E. M. Lifshitz, *The Classical Theory of Fields* (Cambridge, Mass.: Addison-Wesley, 1951), p. 247.

[269] A. Einstein, *Morgan Manuscript*, EA 2070.

[...] according to the general theory of relativity even empty space (empty in the said sense) has physical qualities, which are characterised mathematically by the components of the gravitational potential. [270]

He stated much the same thing in the *Morgan Manuscript*:

Thus, once again "empty" space appears as endowed with physical properties, *i.e.*, no longer as physically empty, as seemed to be the case according to special relativity. One can thus say that the ether is resurrected in the general theory of relativity, though in a more sublimated form.[271] **d52**

The following is another example of how significantly Einstein's epistemological views changed. In 1916, presenting his epistemological argument for the non-existence of space and time in his *Essentials of the General Theory of Relativity*, Einstein said that the generalised principle of relativity deprived time and space of whatever little objective reality they had. And only four years later, in his discussion with Lenard in Bad Nauheim, the same principle was used by Einstein to define the new ether identified with the space-time continuum and treated as something absolutely real.

Einstein: As a matter of principle there is no system of coordinates preferred because of its simplicity [...] The starting point is not an affirmation, but a new definition of the term "ether."[272] **d53**

Einstein's views on the relations between the space-time continuum and matter consisting of particles changed quite fundamentally in the period of 1916-1924. Initially he treated the space-time continuum as something secondary to matter, whereas he later treated space-time as something primary to matter. We can document this evolution with a few quotations. A passage from Einstein's 1918 article criticising Willem de Sitter's work, in which space-time could be considered in the General Theory of Relativity as something real without the presence of any matter, proves that at the beginning Einstein treated the four-dimensional space-time continuum as something secondary to matter.

In my opinion, the general theory of relativity constitutes a satisfying system only if it describes the physical properties of space as completely defined by the presence of matter. Therefore, there cannot

---

[270] A. Einstein, "Dialog über Einwande...," p. 702.

[271] A. Einstein, (*Morgan Manuscript*) par. 222.

[272] "Allgemeine Diskussion über Relativitätstheorie," p. 667.

possibly be a field $g_{\mu\nu}$, that is to say, no space-time continuum, without matter producing it.[273] **d54**

In the same month and year (*i.e.*, in June 1918), in his work "Fundamentals of the General Theory of Relativity,"[274] Einstein gave a new definition of Mach's principle, repeating the same idea:

> *Mach's principle*: the field $g_{\mu\nu}$ is *entirely* determined by the masses of the bodies. Given that according to the results of the special theory of relativity mass and energy are the same thing, and that energy is described formally by means of the symmetric energy tensor $T_{\mu\nu}$, one can deduce that the field $g_{\mu\nu}$ is generated and defined by the energy tensor of matter.[275] **d55**

As we can see, even though he had given up the non-existence of space and time, Einstein did not give up Mach's principle. According to the new definition of that principle, the real four-dimensional continuum equipped with the metrics $g_{\mu\nu}$ was activated by all the masses distributed in the universe.

Less than a year later, in the first half of 1919, a quite significant change in Einstein's views on the relationship between the new ether (*i.e.* the $g_{\mu\nu}$ field) and matter took place. In his work entitled "Do Gravitational Fields Play an Essential Part in the Structure of the Elementary Particles of Matter,"[276] Einstein did, to a certain extent, take into consideration the primary character of the field $g_{\mu\nu}$ in relation to the matter consisting of particles. Einstein came to the conclusion that the gravitational fields could play a role in the formation of these particles. According to Einstein, the electromagnetic fields also contributed to that formation process.

> The previous reflections indicate the possibility of a theoretical construction of matter starting exclusively from the gravitational and electromagnetic fields [...][277] **d56**

Therefore, it is easily understood that the following sentence could appear in Einstein's works on the new ether published at that time:

---

[273] A. Einstein, "Kritisches zu einer von Hrn. de Sitter gegebenen Lösung der Gravitationsgleichungen," *SPAW*, 1 Tl. (1918), pp. 270-272.

[274] A. Einstein, "Prinzipielles zur allgemeinen Relativitätstheorie," *AdP*, 55 (1918), pp. 241-244.

[275] *Ibid.*

[276] A. Einstein, "Spielen Gravitationsfelder im Aufbau der materiellen Elementarteilchen eine wesentliche Rolle?," *SPAW*, 1. Tl. (1919), pp. 349-356.

[277] *Ibid.*

It does remain allowed, as before, to imagine a space filling medium and to assume the electromagnetic fields (and matter as well) are its states.[278] **d57**

In 1924, he dreamt of a theory of fields such that the concept of ether would include "all objects of physics," because according to a consistent field theory, ponderable matter, or the elementary particles it consists of, should be understood as "fields of a special kind," or as "special states of space."[279]

It was not until the later period, when he himself attempted to formulate a unified field theory, that he would write the following words:

> We have now come to the conclusion that space is the primary thing and matter only secondary; we may say that space, in revenge for its former inferior position, is now eating up matter.[280]

What a fundamental change in Einstein's views! Having started from denial of the existence of space and time, he finally came to the conclusion that four-dimensional space (the space-time continuum) constitutes a reality ontologically primary even to matter. At this point, he believed that matter was born from space-time.

---

[278] A. Einstein, *Morgan Manuscript*, section 13.

[279] A. Einstein, "Über den Äther," p. 85.

[280] A. Einstein, "The Concept of Space," *Nature*, 125 (1930), pp. 897-898.

Chapter 4

# DEVELOPMENT OF EINSTEIN'S ETHER CONCEPT (1925-1955)

The further development of the new ether concept is connected with Einstein's attempts to create a relativistic unified field theory. As we know, a unification of gravitational and electromagnetic interactions was the objective of the theory. The General Theory of Relativity ether, described, in terms of structure, by the symmetric metric tensor $g_{\mu\nu}$ was only a "gravitational ether," as Einstein called it in his lecture in Leiden. The General Theory of Relativity ether determined only the inertial-gravitational behaviour of elementary particles. In the General Theory of Relativity, the electromagnetic field is "something in the space-time continuum" but does not belong to its structure. Einstein believed, however, that the cause of this situation was the deficiencies of our theoretical constructions and not physical reality. In his view, the real relativistic ether was the medium for both interactions. As we know, the first attempts to introduce a relativistic ether transmitting both interactions were made by Weyl and Eddington. Kaluza also sought a unification of the two interactions in a five-dimensional continuum. Up to 1925, Einstein had only studied those attempts and critically examined the possibilities they offered. He was not satisfied with any of them, however, and this is why he started looking for his own way to solve the problem of unification.

Ludwik Kostro, *Einstein and the Ether* (Montreal: Apeiron, 2000).

## 4.1 Einstein's first attempt to solve the unification problem

His first attempt was the result of two years of work. Publishing it in 1925 in an article entitled "Unified Field Theory of Gravitation and Electricity,"[281] Einstein was convinced that he had found a final and satisfying solution to the problem. He stated this at the very beginning of the article:

> After a continuous search during the last two years, I believe now that I have found the true solution. I present it in the following.[282] **e1**

What did the solution consist of? In short, it consisted in modifying the mathematical formula of the General Theory of Relativity in such a way that the metric tensor $g_{\mu\nu}$, which was assumed to be a symmetric tensor in the theory as it existed, such that it met the condition $g_{\mu\nu} = g_{\nu\mu}$, ceased to be symmetric, that is $g_{\mu\nu} \neq g_{\nu\mu}$. The existing tensor had only ten components because $g_{\mu\nu} = g_{\nu\mu}$. Since in the new tensor $g_{\mu\nu} \neq g_{\nu\mu}$, the number of components became sixteen.[283]

The ten components of the tensor used previously fully described the gravitational field without taking the electromagnetic field into account. Einstein believed the asymmetric part of the new tensor (the additional six components) described the electromagnetic field. To be more precise, Einstein identified the tensor of the electromagnetic field intensity $F_{\mu\nu}$ with the antisymmetric part of the $g_{\mu\nu}$ tensor, while its symmetric part still described the gravitational field. We can therefore say that in this attempt at unification, the relativistic ether, in addition to a gravitational ether, also became an electromagnetic ether. One might think that Einstein had achieved his goal and that he should have written an article on the new, broader concept of a relativistic ether. Nothing of the sort happened, however, because Einstein was satisfied with this solution for a short time only, and then abandoned it for quite a long time. He would return to it in the last years of his life.

---

[281] A. Einstein, "Einheitliche Feldtheorie von Gravitation und Elektrizität," *SPAW (pmK)* (1925), pp. 414-419.

[282] *Ibid.*, p. 414.

[283] H. Arodź, "A. Einsteina problem unifikacji oddziaływań fundamentalnych," *PF*, 37 (1986), p. 306.

## 4.2     The Kaluza-Klein pentadimensional continuum

After abandoning his 1925 concept, Einstein again turned to Kaluza's solution. He presented his new ideas in 1927 in an article[284] that began with the following words:

> Here the results of further reflections will be reported, whose consequences seem to me to speak strongly in favour of Kaluza's ideas.[285] **e2**

Kaluza's model was significantly improved by Einstein, who did not know that similar improvements had been made by Klein a year before. Today, the improved version of Kaluza's model is called the Kaluza-Klein model. While writing this paper, Einstein thought that his improvements gave a quite satisfactory solution to the unification problem, and consequently he ended the article with the following words:

> Summarising, one can say that Kaluza's approach offers a rational foundation for Maxwell's electromagnetic equations in the framework of the general theory of relativity and that it brings them to a formal unification with the gravitational equations.[286] **e3**

Einstein did not publish any articles on the relativistic ether after this piece. Yet a veritable torrent of articles on the new ether, space and field was to emerge in connection with the next attempt at unification.

## 4.3     Space-time continuum with teleparallelism

Einstein's next attempt at formulating a unified field theory was fully geometric. It resulted in a new version of the relativistic ether—four dimensional space-time continuum with teleparallelism. It is possible to compare the directions of vectors fixed at different points of this space-time continuum. A comparison of this kind was impossible in Riemann's geometry because comparing the directions of vectors required a translation of the vector, and the result of this translation depended on the way it was performed. Einstein would interpret his new geometric theory in its four-dimensional version as a unified field theory, bringing together the gravitational and electromagnetic interactions.[287]

Einstein's first attempt to enrich Riemann's geometry with teleparallelism (also called absolute parallelism) was of a purely

---

[284] A. Einstein, "Zu Kaluzas Theorie des Zusammenhanges von Gravitation und Elektrizität," *SPAW (pmK)*, 1927, pp. 23-30.

[285] *Ibid.*, p. 23.

[286] *Ibid.*, p. 30.

[287] H. Arodź, *op. cit.*, p. 306.

mathematical character. It presented only a new $n$-dimensional geometry and did not include any physical interpretation. Einstein published it in June 1928 under the title "Riemann Geometry with Retention of the Concept of Teleparallelism."[288] Its results were later popularised in some of Einstein's papers on the new version of the ether. When he introduced his new geometry, Einstein did not know that he was following a path that had been blazed before. It turned out that, beginning in 1929, this sort of geometry was introduced and developed by two mathematicians: Elie Cartan and Weitzenblock. Einstein recognised the contribution of both mathematicians and emphasised its significance for his unitary field theory in his paper "On the Unitary Field Theory based on the Riemann Metric and Teleparalellism,"[289] published at the end of 1929. He also thanked Cartan for publishing the history of the new geometry.

On June 14, 1929, in a few papers presented at the Thursday meetings of the Prussian Academy of Sciences, Einstein used the new geometry to develop a unitary field theory, using the new geometry with $n = 4$, of course. In the preface to his work entitled "New Possibilities for a Unitary Field Theory of Gravitation and Electricity,"[290] he wrote:

> Since then I discovered that this theory—at least to a first approximation—leads very simply and naturally to the laws of the gravitational and electromagnetic fields. For this reason it is thinkable that this theory will replace the original version of the general theory of relativity.[291] **e4**

Einstein encountered some mathematical difficulties with the field equations. He did not find any simple and unambiguous way of deriving these equations from Hamilton's principle, and so sought other ways. In a paper entitled "On the Unitary Field Theory,"[292] presented on January 10, 1929, he put forward the equations which he had managed to introduce, as he himself admitted, "in a different way." He was convinced that he had succeeded in deriving the field equations "to a first approximation." Einstein stated:

---

[288] A. Einstein, "Riemanngeometrie mit Aufrechterhaltung des Begriffes des Fern-Parallelismus," *SPAW (pmK)* (1928), pp. 217-221.

[289] A. Einstein, "Auf die Riemann-Metrik und den Fern-Parallelismus gegrundete einheitliche Feldtheorie," *MA*, 102 (1930), pp. 685-697.

[290] A. Einstein, "Neue Möglichkeiten für eine einheitliche Feldtheorie von Gravitation und Elektrizität," *SPAW (pmK)* (1928), pp. 224-227.

[291] *Ibid.*, p. 224.

[292] A. Einstein, "Zur einheitlichen Feldtheorie," *SPAW (pmK)* (1929), pp. 2-7.

More profound research into the consequences of the field equations [...] must show whether the Riemannian metrics, together with the requirement of teleparallelism, really produces an adequate description of the physical properties of space. In the light of the present research this is not improbable.[293] **e5**

Two physicists, Cornelius Lanczos and Hermann Muntz, expressed their doubts about accepting field equations that had not been unambiguously derived from Hamilton's principle. Einstein therefore worked hard to find a rigorous derivation of the field equations from the Hamilton principle. He presented the positive result of his efforts on March 21, 1929, in his paper "Unitary Field Theory and Hamilton's Principle."[294] In the preface to that work, he wrote:

[...] I found that it is possible to solve the problem in a completely satisfactory way, basing the calculations on a Hamilton principle.[295] **e6**

Because he had managed to derive field equations from Hamilton's principle, until 1931 he remained convinced that he had developed a fully satisfactory unitary field theory, unifying the gravitational and electromagnetic interactions.

Einstein's works on the new unitary field theory received a great deal of publicity. It was mainly the daily newspapers that thought he was making another historic discovery. One of his articles, full of mathematical formulas, was hung in the display window of a London department store. Eddington, who witnessed the event, told Einstein that there had been crowds trying to read and, perhaps, understand the article. At first, Einstein himself began popularising his ideas in large-circulation daily newspapers. After some time, however, he started to avoid journalists.[296]

On January 26, 1929, the *Daily Chronicle* published an interview with Einstein, who expressed his belief that he had found the proper solution to the unification problem. Some passages from that article were also published in *Nature* magazine. In it Einstein said:

I believe now that I have found a proper form [...][297] **e7**

---

[293] *Ibid.*, p. 7.

[294] A. Einstein, "Einheitliche Feldtheorie und Hamiltonsches Prinzip," *SPAW (pmK)* (1929), pp. 156-159.

[295] *Ibid.*, p. 156.

[296] H. Arodź, *op. cit.*, pp. 306-307.

[297] A. Einstein, *Nature*, 2/2/1929, p. 175.

On February 3, 1929, he published his article "Field Theories, Old and New"[298] in the *New York Times*. On February 4, 1929, the same article with a slightly modified title, "The New Field Theory," was published in the *Times* newspaper in London.[299] The same article, divided into two parts, was published again in 1930 in *Observatory*.[300] In the article, Einstein shows how Newton's absolute and unchangeable space, whose reality had been assumed to explain the reality of inertia and acceleration, was transformed, not without the use of the field concept introduced by Faraday and developed by Maxwell, into a unitary field theory, which is the third stage of the development of the theory of relativity. As a result of these deep transformations, it had to be admitted that physical space could no longer be treated as a neutral, inactive container, as it was originally thought to be.

> Mere empty space was not admitted as a carrier of physical changes and processes. It was only, one might say, the stage on which the drama of material happenings was played. Consequently Newton dealt with the fact that light is propagated in empty space by making the hypothesis that light also consists of material particles [...].

> This was brought about by the Huygens-Young-Fresnel wave theory of light, which the facts of interference and diffraction forced on stubbornly resisting physicists. The great range of phenomena which could be calculated and predicted to the finest detail by the use of this theory delighted physicists and filled many fat and learned books. No wonder, then, that the learned men failed to notice the crack which this theory made in the statue of their eternal goddess. For in fact this theory upset the view that everything real can be conceived as the motion of particles in space. Light waves were, after all, nothing more than undulatory states of empty space, and space thus gave up its passive *rôle* as a mere stage for physical events. The aether hypothesis patched up the crack and made it invisible. The aether was invented, penetrating everything filling the whole space, and was admitted as a new kind of matter. Thus it was overlooked that by this procedure space itself had been brought to life. It is clear that this had really happened, since the aether was considered to be a sort of matter which could nowhere be removed. It was thus to some degree identical with space itself, *i.e.*, something necessarily given with space. Light was thus viewed as a dynamical process undergone, as it were, by space itself. In this way the field theory was born as an illegitimate

---

[298] A. Einstein, "Field Theories, Old and New," *New York Times*, 3/2/1929.

[299] A. Einstein, "The New Field Theory," *Times*, London, 4/2/1929.

[300] A. Einstein, "The New Field Theory," *Observatory*, 52 (1930), pp. 82-87 and 114-118.

child of Newtonian physics, though it was cleverly passed off at first as legitimate.[301] **e8**

The concept of a field, first introduced into physics by Faraday was—according to Einstein's article—the next step in discovering the active role of space. Further development of the field concept and the demonstration of the active role of space was accomplished—according to Einstein—in the theory of relativity, which (starting in February 1929) "has been in its third stage for the past 6 months."[302]

The above words indicate that Einstein was convinced that he was making an historic discovery. Einstein also said that he had obtained his results using the method of increasing formal simplicity and deduction, *i.e.*, a method very different from the "inductionism" of the positivists.

> Advances in scientific knowledge must bring about the result that an increase in formal simplicity can only be won at the cost of an increased distance or gap between the fundamental hypotheses of the theory on the one hand, and the directly observable facts on the other. Theory is compelled to pass more and more from the inductive to the deductive method, even though the most important demand to be made of every scientific theory will always remain that it must fit the facts.[303] **e9**

Einstein did not mention the new ether at all in this article. He would do so, however, in December 1929, in a series of lectures and articles, in which he also presented the new version of the ether. It is not surprising that, in this version, the four-dimensional pseudo-Riemannian space enriched by teleparallelism functions as an ether transmitting gravitational and electromagnetic interactions.

In December 1929, during the second winter session of the Science Support Association named after Emperor Wilhelm, held in Goethe's Hall in the new Harnack House in Berlin-Dahlem, Einstein gave a lecture entitled "Space, Ether and Field in Physics." The press reported[304] that the lecture was attended by a large audience. At that time Einstein was the director of the Institute of Physics named after Emperor Wilhelm. The *Deutsche Bergwerks-Zeitung*[305] published a thorough summary of the

---

[301] *Ibid.*, pp. 83-84.

[302] *Ibid.*, p. 114.

[303] *Ibid.*, pp. 114-115.

[304] [Anonymous], "Professor Einstein spricht über das physikalische Raum- und Äther-Problem," *Deutsche Bergwerks-Zeitung*, 15/12/1929, p. 11.

[305] *Ibid.*

main ideas of the lecture. The whole lecture was published in the magazine *Forum Philosophicum*,[306] along with an English translation.[307]

An article of similar content and not very different title ("The Problem of Space, Field and Ether in Physics") was published in *Die Koralle*,[308] a monthly magazine for admirers of nature and technique. On June 7, 1939, during his stay in England, at the University of Nottingham, Einstein gave a lecture on the same subject. The lecture was translated into English by Dr. I. H. Brose. The *New York Times* obtained the short-hand minutes of the translation for publication. The report was also published, with the permission of the newspaper's editors, in the magazine *Science*.[309] The main ideas of the lecture, based on notes by H.T.H. Piaggio, were published in *Nature*.[310]

An almost identical lecture was given by Einstein during the Second World Power Conference, held in Berlin on June 16-25, 1930. The first conference had been held in London in 1924. According to E. Kuhn,[311] who wrote a report on the Berlin conference, the conference attracted an audience of over 3900 participants from forty-eight countries. There were 380 lectures on specific topics and thirty-four on general topics. All the lectures were published in twenty-one volumes. According to Kuhn, the introduction of the so-called leading lectures, given by the most distinguished scientists representing many countries, was the novelty of the conference.

The series of leading lectures started with Einstein's lecture entitled "The Problem of Space, Field and Ether in Physics."[312] The second leading lecture, entitled "Subatomic Energy,"[313] was delivered by Eddington, who, based on Einstein's equation $E = mc^2$, showed where the Sun and stars obtained their energy. Einstein's opening lecture conveyed a message of its own. Because he had developed a unitary field theory, Einstein was convinced that physical space, closely tied to time as

[306] A. Einstein, "Raum, Äther und Feld in der Physik," *FPh*, 1 (1930), pp. 173-180.

[307] A. Einstein, Space, "Ether and Field in Physics," *FPh*, 1 (1930), pp. 180-184.

[308] A. Einstein, "Das Raum-, Field- und Äther-Problem in der Physik," [cited here as "Das Raum-..."] *Die Koralle*, 5 (1930), H. 11, pp. 486-488.

[309] A. Einstein, "Address at the University of Nottingham," *Science*, 71 (1930), pp. 608-610.

[310] H.T.H. Piaggio, "The Concept of Space," *Nature*, 125 (1930), pp. 897-898.

[311] E. Kuhn, "Zweite Weltkraftkonferenz Berlin 1930," *Dinglers polytechnisches Journal*, 345 (1930), H. 7, pp. 121-123.

[312] A. Einstein, "Das Raum-, Field- und Äther-Problem in der Physik," [cited here as: "Das Raum-, Feld-..."] *Transactions of the Second World Power Conference*, 19 (1930), pp. 1-5.

[313] E. Kuhn, *op. cit.*, p. 123.

a total field of gravitational and electromagnetic interactions, constituted the most basic source of energy. This was why he thought it was a good idea to open a conference on energy resources with a lecture on this subject.

The lecture series also included the article entitled "The Problem of Space, Ether and Field as a Problem of Physics."[314] It was published in 1934 in a collection of Einstein's speeches and articles entitled *Mein Weltbild*. Yet by 1934, Einstein had already viewed his geometric unified field theory as unsuccessful for three years.

In the above mentioned works, Einstein presented his new version of relativistic ether. Here, the ether is identified with physical space understood as a pseudo-Riemannian four-dimensional continuum, enriched by teleparallelism. Gravitational and electromagnetic fields constitute states of the continuum. They belong to its structure and are dependent on it. Thus, electromagnetic fields ceased to be "something in the space-time continuum" and became totally dependent on its structure. In the new version, the ether cannot be assigned any state of motion, either. In his article published in *Mein Weltbild*, Einstein wrote:

> Physical space and ether are only different terms for the same thing; fields are physical states of space. If no particular state of motion can be ascribed to the ether, there do not seem to be any grounds for introducing it as an entity of a special sort alongside space.[315] **e10**

Of the three terms, "ether," "physical space" and "field," which at that time had become synonyms in the theory of relativity, in his works on relativistic ether published between 1918 and 1924, Einstein used the term "ether" most often and believed that in a coherent field theory the term would include all objects of physics. After the development of a geometrically unitary field theory, in Einstein's work the term "physical space" was used more frequently. According to Einstein, it was "physical space" that took over the functions of the ether, which had been introduced because scientists did not want to ascribe variability and physical activity to it. In the new field theory, the concept of four-dimensional physical space, which possesses metric and direction (teleparallelism), is to include all of reality.

> The real is conceived as a four-dimensional continuum with a unitary structure of a definite kind (metric and direction). The laws are

---

[314] A. Einstein, "Das Raum-, Äther- und Feld-Problem der Physik" in: A. Einstein, *Mein Weltbild* (Amsterdam: Querido, 1934), pp. 229-248.

[315] *Ibid.*, p. 237.

differential equations, which the above structure satisfies, namely, the fields that appear as gravitation and electromagnetism. The material particles are positions of high density without singularity.

We may summarise in symbolical language. Space, brought to light by the corporeal objects and made a physical reality by Newton, has in the last few decades swallowed ether and time and also seems about to swallow the field and the corpuscles, so that it remains as the sole carrier of reality.[316] **e11**

For Einstein, four-dimensional physical space, fulfilling the functions of the ether, was something primary to matter consisting of particles. During the lecture at the University of Nottingham, Einstein would say:

The strange conclusion to which we have come is this—that now it appears that space will have to be regarded as a primary thing and that matter is derived from it, so to speak, as a secondary result. We have always regarded matter as a primary thing and space as a secondary result. Space is now having its revenge, so to speak, and is eating up matter. But that is still a pious wish.[317] **e12**

Einstein added the last words because up to 1930 he had not managed to find solutions of the field equations (singularity-free solutions) that could be interpreted as a mathematical representation of elementary particles. As we will see, later (in 1935) Einstein claimed that he had found such solutions representing particles with and without electrical charge.

Having come to the conclusion that geometry had again provided him with a proper tool to solve the problem, this time the problem of unification, Einstein started to work on the origin of the concept of space, which plays a key role in geometry. All the works studied at the beginning of this chapter deal mainly with the origin and history of the concept of space. They start with a handful of remarks on the role of concepts in human cognition in general. According to Einstein:

Concepts and conceptual systems always are useful in organising our experiences and giving them "sense."[318] **e13**

Concepts, however, cannot in any way be logically derived from the experience of our senses, although, ultimately, they always have to be related to them.

---

[316] A. Einstein, "Raum, Äther...," pp. 179-180.

[317] A. Einstein, "Address at the University of Nottingham," p. 610.

[318] A. Einstein, "Das Raum-...," p. 486.

We may answer that concepts, considered logically, never originate in experience; i.e., they are not to be derived from experience alone. And yet they are formed in our mind only with reference to what is experienced by the senses. We have to explain such fundamental concepts by pointing out the characteristic of our sense-experiences which has led to the formation of the concept[319]. **e14**

It is easy to identify sensations in the case of concepts directly connected with senses. It becomes much more difficult, however, with abstract concepts, which, albeit indirectly, are also related to sensations.[320] The concept of space, which Einstein considered to be an abstract notion, was closely connected to the idea of the physical object. Together with the concept of relations between objects, it was considered to anticipate the concept of space.

> From this point of view it becomes now apparent that the concept of space is linked with the concept of material object. If this ideal description is followed, those complexes of experiences which we describe abstractly as "positions of bodily objects" turn out to be particularly simple. It is clear that the positional relationships of bodies are real in the same sense as the bodies themselves.[321] **e15**

> Accordingly: without the concept of body, no concept of spatial relation among bodies; and without the concept of spatial relation, no concept of space.[322] **e16**

In pre-scientific cognition, space was one universal quasi-body, omnipresent, all-pervading, in contact with everything, born in our minds for us to be able to understand the distribution of objects.

> When considering the mutual relations of the location of bodies, the human mind finds it much simpler to relate the location of all bodies to that of a single one rather than to grasp mentally the confusing complexity of the relations of every body to all others. This one body, which is everywhere and must be capable of being penetrated by all others in order to be in contact with all, is indeed not given to us by the senses, but we devise it as a fiction for convenience in thought.[323]

---

[319] A. Einstein, "Raum, Äther...," p. 173.

[320] A. Einstein, "Das Raum-...," p. 486.

[321] *Ibid.*

[322] A. Einstein, "Raum, Äther...," p. 173.

[323] *Ibid.*

Compared to pre-scientific knowledge, the achievements of the ancient Greeks in geometry were a great step forward, although the concept of space as such did not appear in that geometry.

> For in the oldest geometry, which the Greeks gave us, investigation is limited solely to the local relations of idealised corporeal objects, which are called "point," "line," and "plane." In the concepts of "similarity" and "measurement" the reference to the local relations of corporeal objects is plainly shown. A spatial continuum, in short, "Space," is not to be found in the Euclidean geometry at all, in spite of the fact that this concept had of course been current in prescientific thought.

> The extraordinary significance of ancient Greek geometry lies in the fact that it is, as far as we know, the first successful attempt to comprehend a complex of sense-experience conceptually by means of a logically deductive system.[324] **e18**

It was only in modern science that the concept of space emerged. We owe it to Descartes:

> The spatial continuum as such was introduced into geometry by the moderns, first of all by Descartes, the founder of analytic geometry.

> Descartes's service in introducing the spatial continuum into geometry can scarcely be overestimated [...] it decidedly heightened the scientific character of geometry. For the straight line and the plane were henceforth no longer fundamentally privileged above other lines and surfaces, but all lines and surfaces now experienced a similar treatment. The complicated system of axioms in Euclidean geometry was replaced by a single axiom which runs as follows, in contemporary language: There are systems of coordinates in relation to which the distance $ds$ between adjacent points $P$ and $Q$ is expressed in terms of the differences of the coordinates $dx_1$, $dx_2$, $dx_3$ by means of the Pythagorean proposition, i.e., by the formula

$$ds^2 = dx_1^2 + dx_2^2 + dx_3^2$$

> From this, *i.e.*, from the Euclidean metric, all concepts and propositions of Euclidean geometry may be deduced.[325] **e19**

Cartesian co-ordinates were invented to simplify our thinking. They could not be assigned the characteristics of objective reality. It was not until Newton's mechanics that space became something totally real.

---

[324] *Ibid.*, p. 174.
[325] *Ibid.*, pp. 174-175.

According to what has been said up to now the spatial relations of bodies have a physical reality, but not space itself. Space receives a physical reality in Newtonian mechanics, because acceleration appears in it as a fundamental concept in the law of motion. Acceleration is here a state of motion relative to space which cannot be reduced to a position relative other bodies. Metrics and inertia are thus, in Newtonian physics, the most essential properties of space.[326] **e20**

It was then that the mechanistic image of the world, according to which there existed space and time, and material bodies moving in them, was developed. The image started to fall to pieces the moment the concept of field was introduced by Faraday and Maxwell.

In the beginning, the desire to consider the field as a mechanical state of an omnipresent matter, the ether, was dominant. Given that such a point of view did not generate satisfactory progress, one kept to the ether as a special material stuff whose states had to determine the field, but the mechanical interpretation of these states was abandoned.[327] **e21**

The next step was made by Lorentz, who immobilised the ether against space. Then the ether could have been identified with space. It was not because of the deeply rooted conviction about the invariability of space.

Towards the end of the last century, H.A. Lorentz showed that one could not attribute to the ether any global movement with respect to space if electromagnetic processes were to be described in a quantitatively correct way.[328] **e22**

Was it not natural then to say: The fields are states of space; space and ether are the same thing. That it wasn't said was due to this: space as the seat of Euclidean metrics and Galileo-Newton's inertia was considered absolute, that is, not influenceable, almost a rigid skeleton of the world, which, so to speak, pre-exists all physics and which cannot be the carrier of changing states.[329] **e23**

The differentiation between the ether and space disappeared by the very nature of things in the Special Theory of Relativity.

---

[326] A. Einstein, "Das Raum-...," pp. 486-487.

[327] *Ibid.*, p. 487.

[328] *Ibid.*

[329] A. Einstein, "Das Raum-, Feld-...," p. 3.

The separation of the concepts of space and ether has in a certain sense disappeared by itself, after the special theory of relativity took away from the ether the last residue of substantiality.[330] **e24**

The superstition about the absolute nature of space, *i.e.* its invariability, was rejected in the General Theory of Relativity. Concerning general relativity, Einstein notes:

> In this way space lost its absolute character. It was endowed with variable (in conformity with the laws) states and processes, so that it could assume the functions of the ether, and—as far as the gravitational field was concerned—really assumed them. Only the formal meaning of the electromagnetic field remained provisionally obscure, as it could not be interpreted as a pure metric structure of space.[331] **e25**

> To this end one should also try to find a structure of great formal wealth, which could include the Riemannian metric structure, and which at the same time could mathematically describe the electromagnetic field.[332] **e26**

The new unitary field theory went in the direction in which Riemann's structure of space had been enriched by teleparallelism. In that theory, the enriched four-dimensional space fulfilled the functions of the ether for both the gravitational field and the electromagnetic field. Referring to this, Einstein wrote (we repeat the statement quoted above):

> Physical space and the ether are only different terms for the same thing; fields are physical states of space.[333] **e27**

This, in short, is the history of the concept of space in Einstein's lectures and works mentioned above. It starts with the concept of physical object and ends with the concept of the four-dimensional space continuum, which provides for the gravitational and electromagnetic interactions of the ether.

In the second half of 1930 and the first months of 1931, Einstein, alone or in collaboration with Walther Mayer, tried to improve the new geometry and the related unitary field theory[334] in a few works. Einstein

---

[330] *Ibid.*, p. 4.

[331] A. Einstein, "Das Raum-...," p. 487.

[332] A. Einstein, "Das Raum-, Feld-...," p. 5.

[333] A. Einstein, "Das Raum-, Äther...," p. 237.

[334] A. Einstein, "Die Kompatibilität der Feldgleichungen in der einheitlichen Feldtheorie," *SPAW (pmK)* (1930), pp.18-23; A. Einstein, W. Mayer, "Zwei strenge statische Lösungen der Feldgleichungen der einheitlichen Feldtheorie," *SPAW (pmK)*

mentioned the results in one of his lectures in England. In May 1931, Einstein was staying in Oxford, where, at Rhodes House, he delivered three lectures under the title "Theory of Relativity: Its Formal Content and its Present Problems."[335] On May 23, he received a Doctorate *Honoris Causa* from Oxford University. On the occasion of that ceremony, he delivered his third lecture at Rhodes House, in which he presented his geometric unified field theory, with the improvements mentioned above. The main ideas of the lecture, like the two previous lectures, were published as a note in *Nature*.[336] We learn from the note that Einstein also suggested that his new theory might also have some influence on quantum physics. This means that in May of 1931 he was still convinced that his geometric unitary field theory was fully correct. Soon afterwards, however, he decided it was unsatisfactory, and abandoned it forever.

## 4.4    Four-dimensional space-time with pentavectors

A note on Einstein's and Mayer's new attempt to develop a unitary field theory can be found in *Science* of October 30, 1931.[337] Einstein announced the publication of the new attempt in "the nearest future." With Einstein's permission, *Science* also published a short preliminary presentation of the new theory, which he had prepared for the Josiah Macy, Jr. Foundation, which financed the work. A passage from the first part of the presentation reads as follows:

> After we both had worked more than a year on the further development of the last theory, we reached the conclusion that we were striving in the wrong direction and that the theory of Kaluza, while not acceptable, was nevertheless nearer the truth than the other theoretical approaches.[338] **e28**

As we already know, Kaluza's theory assumes that the space-time continuum is pentadimensional. Assuming that Riemann's metric exists

---

(1930), pp. 110-120; A. Einstein, "Zur Theorie der Räume mit Riemann-Metrik und Fernparallelismus," *SPAW (pmK)* (1930), pp. 401-402; A. Einstein, W. Mayer, "Systematische Untersuchung über kompatible Feldgleichungen welche in einen Riemannschen Raume mit Fernparallelismus gesetzt werden können," *SPAW (pmK)* (1931), pp. 257-265.

[335] A. Einstein, "Theory of Relativity: Its Formal Content and its Present Problems, a short presentation," in: *Nature*, 127 (1931), pp. 765, 790, 826-827.

[336] *Ibid.*, p. 827.

[337] A. Einstein, "Gravitational and Electrical Fields (Preliminary report for the Josiah Macy, Jr. Foundation)," *Science*, 74 (1931), pp. 438-439.

[338] *Ibid.*, p. 438.

in five dimensions, it contains field laws which, in the first approximation, are in agreement with the laws of gravitation and electromagnetism. Einstein did not like the pentadimensional nature of Kaluza's theory: His main objection was that its five dimensions were inconsistent with the real four-dimensional character of the space-time continuum. To deal with this incoherence, Einstein and Mayer decided to introduce five-vectors and five-tensors in the four-dimensional space-time continuum.

> We have succeeded in formulating a theory which formally approximates Kaluza's theory without being exposed to the objection just stated. This is accomplished by the introduction of an entirely new mathematical concept which may be described as follows: Until now it has been believed that one can introduce into a space of n dimensions only vectors or vector-fields of which the number of components agrees with the number of dimensions of that space. It appears, however, that this restriction is not necessary. It has its origin in the *anschauliche* significance of those vectors responsible for the formulation of the vector concept. We have been successful in introducing into space $R_n$ of $n$ dimensions, vectors $a^i$ $(i = 1... $ m) of m components and in deriving a calculus of such vectors and tensors which is essentially no more complicated than the well-known absolute calculus.
>
> Our theory arises quite readily from consideration of five-vectors (five components) in the four-dimensional continuum. There follows from that a "five-curvature" of space which is analogous to the Riemannian curvature, and which bears a similar relationship to the laws of the unitary field that the Riemannian curvature does to the relativistic equations of the gravitational field alone. This theory does not yet contain the conclusions of the quantum theory. It furnishes, however, clues to a natural development, from which we may anticipate further results in this direction. In any event, the results thus far obtained represent a definitive advance in knowledge of the structure of physical space.[339] **e29**

As we can see, Einstein assured his readers that he had made a significant step forward in divulging the structure of physical space. Many physicists stopped taking Einstein's claims seriously. The first presentation of the new theory developed jointly with Mayer was made by Einstein on October 22, 1931, at the meeting of the Prussian Academy of Sciences. The presentation was entitled "Unitary Theory of

---

[339] *Ibid.*, p. 439.

Gravitation and Electricity."[340] In the preface, Einstein again strongly emphasised both authors' belief about the decisive significance of the new theory:

> [...] we want to propose here a theory which, apart from the quantum problem, we believe gives a completely satisfactory final solution.[341]
> **e30**

The term "consistent unitary theory of total field" also appeared for the first time in the preface. From that time on, the term "total field" (*Gesamtfeld*) would become a synonym for the terms "physical space" and "the ether," and in time would play a leading role. The term "total field" was to become Einstein's favourite term, and the term "ether" would not be used as often as before.

Einstein dreamt of finding solutions of the total field equations that could represent elementary particles. He admitted, however, that the authors did not manage to achieve this result in the present theory.[342]

The second paper prepared jointly by Einstein and Mayer,[343] with the same title as the first, sought to depict the structure of space in such a way that the essence of material particles could be represented in a natural way. The authors stressed this in the preface:

> It is our conviction that a satisfactory field theory must be in agreement with a singularity-free description of the total field, and therefore also of the field inside the corpuscles. For this reason we asked ourselves the question if the structure of space which we considered might not permit a generalisation leading to electromagnetic equations with an electric density different from zero. In the following it will be shown that such a generalisation exists, that it is very natural, and that it affords the opportunity to formulate a compatible system of field equations. The question of the adequacy of this system of equations for the description of reality will not be dealt with here.[344] **e31**

The paper did not present a final field solution that could represent particles. The solutions that Einstein found satisfactory were to be published, in co-operation with Nathan Rosen, in 1935—but that would

---

[340] A. Einstein, W. Mayer, "Einheitliche Theorie von Gravitation und Elektrizität," *SPAW (pmK)* (1931), pp. 541-557.

[341] *Ibid.*, p. 541.

[342] *Ibid.*, p. 556.

[343] A. Einstein, W. Mayer, "Einheitliche Theorie von Gravitation und Elektrizität," 2 Abhandlung, *SPAW (pmK)* (1932), pp. 130-137.

[344] *Ibid.*, p. 130.

be in another context and in another place, *i.e.*, not as part of the total field theory with five-vectors discussed here, and not in Germany, but after Einstein had moved to the United States.

## 4.5    Anti-Einstein campaign. Einstein leaves Europe

Let us go back to the year 1920, the year in which the famous debate between Lenard and Einstein took place in Bad Nauheim. After that debate, the anti-Einstein campaign organised by the founders of the so-called German school of physics (Lenard, Gehrcke and others) did not stop. Quite the opposite: it became more vigorous. The problem of the ether still furnished the backdrop for that campaign.

In 1921, Lenard published his book entitled *Ether and Para-ether.*[345] Though not admitted publicly, the main purpose of the book was to attack Einstein and his theory. Lenard did everything he could to challenge Einstein's authority and to ridicule his theory. In 1922, the second, revised edition of the book was published.[346] It was published deliberately (according to Gehrcke[347]), shortly before the 1922 conference of German natural scientists in Leipzig. This can be clearly seen in the preface to the second edition, "A Warning to German Scientists."[348] According to Gehrcke, who published a review[349] of the second edition in 1923, the book was meant to be a protest against lectures on the theory of relativity which were to be delivered at the Conference in Leipzig. Lenard himself presented the objective of his book in the following way:

> As can be seen from the above, in this paper I certainly did not intend to criticise the "theory of relativity"—the natural scientist should have better things to do than criticise this "theory." Rather, what will be reported here as most important is a new way of understanding the processes of the ether, through which the theory of relativity becomes superfluous.[350] **e32**

In the preface, Lenard warned German natural scientists against being influenced by the theory of relativity because it was only a "pile of

---

[345] Ph. Lenard, *Über Äther und Uräther* (Leipzig: Verlag von S. Hirzel, 1921).

[346] Ph. Lenard, *Über Äther und Uräther*, 2 verm. Aufl. (Leipzig: Verlag von S. Hirzel, 1922).

[347] E. Gehrcke, [review] "Ph. Lenard, *Über Äther und Uräther*," 2. verm. Aufl. *ZftP*, 4 (1923), nr 9, p. 334.

[348] Ph. Lenard, *Mahnwort zu Deutsche Naturforscher*, in: Ph. Lenard, *Über Äther und Uräther*, 2 verm. Aufl., pp. 5-10.

[349] E. Gehrcke, [review]..., p. 334.

[350] Ph. Lenard, *Mahnwort*..., p. 5.

hypotheses" (*ein Hypothesenhaufen*), which had been prematurely called a theory.

> Is it then correct to call a pile of hypotheses a "theory"—however mathematically well constructed it may be? The announcement of the name "theory of relativity" is even deceitful in light of its current status.[351] **e33**

Another passage from the preface (also quoted by Gehrcke in his review) suggested that Einstein's work should not be taken seriously because he had already abandoned the concept of the ether only to come back to it and discuss its properties later:

> Didn't Mr. Einstein, after having declared the non-existence of the ether with great emphasis, make it the topic of a lecture in which its existence and its hypothetical properties were discussed? If one can discuss about ether, the same can be done with absolute motion; if one can discuss absolute motion, the same can be done, and radically, with truth and the value of a relativistic theory; is this not evident?[352] **e34**

In the "Warning," the "scientific" elements were mixed with nationalistic diatribes, revealing Lenard's Nazism and anti-Semitism. According to Lenard, the theory of relativity, full of hypotheses based on mathematical tricks and not on facts, lacked the native German rigour.

> Where have German carefulness and concreteness gone?[353] **e35**

A new model of the ether (actually, two ethers) is the main subject of Lenard's book. He says that the ether makes the theory of relativity redundant because it accounts for all the effects seemingly explained by the theory. According to Lenard, there are two ethers. He calls the first ether the "para-ether" (*Uräther*) or "meta-ether" (*Metäther*). It fills all of space and is at total rest. It is a medium which conditions the properties of electromagnetic fields. Electromagnetic waves propagate in it with absolute velocity $c$ in interplanetary space. Apart from the metaether, there is another ether of different density, linked to matter. The more matter there is, the denser the ether surrounding it. Each atom has its own surrounding ether, and that is why celestial bodies consisting of atoms have their own ether in their environment. The ether always accompanies the bodies. The Earth also has an ether of its own. Within

---

[351] *Ibid.*
[352] *Ibid.*, pp. 5-6.
[353] *Ibid.*, p. 6.

this ether, light travels with a velocity which equals *c* relative to the Earth, which is why in the Michelson-Morley experiments the velocity of light travelling in all directions is always the same. Light that leaves the Earth's ether into interplanetary space propagates with absolute velocity *c* in relation to the paraether. By means of the two ethers, Lenard tried to account for some other phenomena, *e.g.*, aberration of light and the Doppler effect. Yet he did so in a qualitative way only, without any mathematical development. The very few mathematical formulas that can be found in his work do not constitute a real development. In light of the above, Lenard's claims that his ethers make the theory of relativity redundant cannot be taken seriously. Lenard's quotation of Weyland's statement that the theory of relativity owes its popularity to mass suggestion brought about by special advertising[354] at the end of the book seems equally ridiculous.

Johannes Stark, a German experimental physicist, joined the attacks against Einstein and his theory. He repeated the objections raised by Lenard and Gehrcke. He further claimed that the theory of relativity was merely a mathematical game which had nothing to do with physical reality. In 1922, he published the article *The Present Crisis in German Physics*,[355] in which he stated that:

> If only Einstein had joined the mathematicians and the philosophers with his new theory from the start! German physics would then perhaps have been saved from the poison that has paralysed its thought, because according to his theory, through analysis based on the clever fiction of "thought experiments" (*Gedankenexperimenten*) and with the aid of mathematical operations it could have obtained a physical understanding or, as it is usually called, a "representation of the world." **e36**

Since Stark treated the existence of the ether as a fact, his critical remarks on Einstein's theory also touched on the problem of the ether:

> Exaggeration in abstraction and formalisms, and self-limitation to an intellectual game with mathematical definitions and formulas can be seen above all in the intentional disregard of the ether. After disregarding ether it is certainly possible to establish physical relations between material bodies by means of mathematical formulas. But when this is done, will the concept of ether become superfluous, will the fact of the existence of ether be thrown out of the world? [...]

---

[354] Ph. Lenard, *Über Äther und Uräther*, 2. verm. Aufl., p. 54.

[355] J. Stark, *Die Gegenwärtige Krisis in der deutschen Physik* (Leipzig: Verlag von J. A. Barth, 1922).

No, the celebrated abolition of the ether by Einstein is not a great accomplishment, but a horrible regression in physical science. The introduction of the ether concept in optics and in electrodynamics and the clear way of thinking so generated have proved extraordinarily productive in physics; the physical research of a century has transformed the ether from an hypothesis to a fact. Physics without the ether isn't physics. Einstein himself now has misgivings about his great achievement of having abolished the ether; in fact, recently he seemed to want to reintroduce the ether in one of his lectures, though certainly not the old abolished ether, but a kind of Einsteinian relativistic ether.[356] **e37**

The problem of the ether was the main battlefield of Gehrcke's repeated attacks on the theory of relativity. In 1923, he published an article "The Contradictions between the Ether Theory and Relativity Theory and Experimental Tests."[357] He made a general analysis of the Michelson-Morley, Michelson-Miller and Sagnac experiments, in which he attempted to show that they allowed a different, better interpretation than that offered by the theory of relativity, based on a modification of the concepts of time and space.

The principle of mass-energy equivalence ($E = mc^2$) is of essential significance in physics. It is one of the most important consequences of the Special Theory of Relativity, in which it finds its full theoretical justification. Lenard and Gehrcke did realise the significance of that principle. That is why they went to great lengths to maintain it without having it associated with Einstein. In his book *On Ether and Para-ether*, Lenard pointed out that an Austrian physicist, Friedrich Hasenöhrl, had discovered that energy has mass. Since Hasenöhrl's equation took the form

$$E = \frac{3}{4}mc^2 .$$

Lenard added his own derivation of the principle, arriving at $E = mc^2$. He did so—as Gehrcke emphasised[358]—without making any reference to the relativity principle.

To undermine Einstein's contribution to the theory of relativity and to the $E = mc^2$ equation, in 1929, Lenard published a book entitled *Great*

---

[356] *Ibid.*, p. 11-12.

[357] E. Gehrcke, "Die Gegensätze zwischen der Äthertheorie und Relativitätstheorie und ihre experimentale Prüfung," *ZftP*, 4 (1923), Nr. 9, pp. 292-299.

[358] *Ibid.*, p. 299.

*Scientists*,[359] intended for the general reader, one long chapter of which was devoted to Hasenöhrl, whom he honoured as one of the most distinguished scientists, for discovering the inertia of energy when experimenting with cathode rays. Also in 1929, another of Lenard's books, *Energy and Gravitation*[360] was published. The book started with the statement:

> Hasenöhrl was the first to demonstrate that energy possesses mass (inertia).[361] **e38**

Lenard, Gehrcke and Stark were not alone in attacking Einstein and his theory. In Germany the anti-Einstein campaign reached such proportions that hundreds of articles attacking his theory of relativity were published. Some authors even produced series of articles. To mention only one, the book *One Hundred Authors against Einstein*,[362] published in 1931, is a collection of speeches and passages from the works of various professors and doctors—opponents of the theory of relativity.

Einstein's life in Germany was becoming more and more unbearable, especially after the Nazis took power on January 30, 1933. On March 20, 1933, when Einstein was away in the United States, his summer house in Caputh was broken into, allegedly to search for arms that were said to have been left there by communists. Eight days later Einstein came back to Europe, but he never returned to Germany. He stayed temporarily in the Savoyarde villa at Le Coq sur Mer in Belgium, protected by two representatives of the Belgian secret police forces. On October 17, 1933, together with his wife, his secretary, Helen Dukas and his co-worker, Walther Mayer, Einstein went to live in the United States, settling at 2 Library Place in Princeton.[363]

Even after Einstein left Germany, the campaign against his theory continued as part of developing Nazism and anti-Semitism. Since the founders of the so-called German physics were at the same time strong supporters of the ether in physics, the scientists and philosophers who were in favour of national socialism kept developing various models of

[359] Ph. Lenard, *Grosse Naturforscher* (Munich: J.F. Lehmanns Verlag, 1929), pp. 308-316.

[360] Ph. Lenard, *Über Energie und Gravitation* (Berlin/Leipzig: Walter de Gruyter und Co., 1929).

[361] *Ibid.*, p. 3.2

[362] H. Israel, E. Ruckhaber, R. Weinmann, *Hundert Authoren gegen Einstein* (Leipzig: R. Voigtlanders Verlag, 1931).

[363] A. Pais, *op. cit.*, p. 592.

ether. There was an astonishing volume of works on the ether. They all pointed to one thing: the theory of relativity was erroneous. We quote one of the extreme examples here. In 1934, Christoph Schrempf published his book, *The Ether as Foundation of a Unified Cosmology*,[364] devoted mainly to the theory of ether vortices. The ether vortex model presented in that book was so "scientific" that it even supported national socialism. The author actually went so far as to state that Mother Nature always creates ether vortices that are swastika-shaped.[365] Given this kind of ether model, we know what to expect in the chapter entitled "The Theory of Ether Vortices *vs.* the Theory of Relativity."[366]

It is no exaggeration to state that the term "ether" always gave the green light for attacks on Einstein and his theory of relativity in Nazi and anti-Semitic Germany. Perhaps this is why Einstein used the term "ether" less and less frequently, preferring "physical space" at first, and later, "total field." However, he still understood the four-dimensional space-time continuum as a special kind of matter, since, because it was a total field, it was a source of energy that was mass-creating, able to create elementary particles.

## 4.6    Elementary particles as "portions" of space

In 1935, in their article entitled "The Particle Problem in the General Theory of Relativity,"[367] Einstein and Rosen presented a new theory, which they regarded as such a generalisation of the General Theory of Relativity (together with Maxwell's theory of electromagnetism) that it provided an atomistic theory of matter and electricity as well. The generalisation consisted in a "minor, insignificant change"—as Einstein put it—in the field equations. The change consisted only in removing the denominator from the equations. It resulted in a very significant effect, however: it removed the singularity appearing both in the General Theory of Relativity and in Maxwell's electromagnetism.

In their own words, here is how Einstein and Rosen eliminated the singularities from the equations of gravitational fields by removing the denominators.

---

[364] Ch. Schrempf, *Der Weltäther als Grundlage eines einheitlichen Weltbildes* (Leipzig: Otto Hillmanns Verlag, 1934).

[365] *Ibid.*, p. 72.

[366] *Ibid.*, p. 78.

[367] A. Einstein, N. Rosen, "The Particle Problem in the General Theory of Relativity," *PhR*, 48 (1935), pp. 73-77.

If in a space free from gravitation a reference system is uniformly accelerated, the reference system can be treated as being "at rest," provided one interprets the condition of the space with respect to it as a homogeneous gravitational field. As is well known the latter is exactly described by the metric field

$$ds^2 = -dx_1^2 - dx_2^2 - dx_3^2 + \alpha^2 x_1^2 dx_4^2 \quad (1)$$

The $g_{\mu\nu}$ of this field satisfy in general the equations

$$R^i_{klm} = 0 \quad (2)$$

and hence the equations

$$R_{kl} = R^m_{klm} = 0 \quad (3)$$

The $g_{\mu\nu}$ corresponding to (1) are regular for all finite points of space-time. Nevertheless one cannot assert that Eqs. (3) are satisfied by (1) for all finite values of $x_1, \dots x_4$. This is due to the fact that the determinant g of the $g_{\mu\nu}$ vanishes for $x_1 = 0$. The contravariant $g^{\mu\nu}$ therefore become infinite and the tensors $R^i_{klm}$ and $R_{kl}$ take on the form $0/0$. From the standpoint of Eqs. (3) the hyperplane $x_1 = 0$ represents a singularity of the field.

We now ask whether the field law of gravitation (and later on the field law of gravitation and electricity) could not be modified in a natural way without essential change so that the solution (1) would satisfy the field equations for all finite points, i.e., also for $x_1 = 0$. W. Mayer has called our attention to the fact that one can make $R^i_{klm}$ and $R_{kl}$ into rational functions of the $g_{\mu\nu}$ and their first two derivatives by multiplying them by suitable powers of $g$. It is easy to show that in $g^2 R_{kl}$ there is no longer any denominator. If we replace (3) by

$$R^*_{kl} = g^2 R_{kl} = 0 \quad (3a)$$

this system of equations is satisfied by (1) at all finite points. This amounts to introducing in place of the $g^{\mu\nu}$ the cofactors $[g_{\mu\nu}]$ of the $g^{\mu\nu}$ in $g$ in order to avoid the occurrence of denominators. One is therefore operating with tensor densities of a suitable weight instead of with tensors. In this way one succeeds in avoiding singularities of that special kind which is characterised by the vanishing of $g_{\mu\nu}$.[368] **e39**

---

[368] *Ibid.*, p. 74.

Some authors believed that particles could be represented as field singularities. According to Einstein and Rosen, however, this standpoint was unacceptable because each singularity brought so many options into the theory that it actually annihilated its laws. When the singularities are removed from the solutions of field equations, the new theory still made use only of the gravitational field variables, which were components of the $g_{\mu\nu}$ tensor (*i.e.*, gravitational potentials) and of the variables of electromagnetic field (in Maxwell's approach), which were vector potentials $\varphi_\mu$, without any new variables added.

In the mathematical formalism of this theory, the four-dimensional space-time continuum was depicted by two identical congruent parts, which were called sheets by the authors of the article. The links between them were called bridges. Such bridges, finite spaces, were interpreted as representations of elementary particles. There were two types of bridge: one representing neutral elementary particles and possessing mass, and the other representing electrically charged particles not possessing mass. Hence, real particles, such as electrons and protons, having both electric charge and mass, were represented by "two-bridges" that connected both sheets of space. Thus, the two kinds of particle (*i.e.*, electrically neutral and electrically charged) constituted portions of space.

The theory was presented in two concrete examples. It made use of the well-known Schwarzschild solutions of the gravitational field for the static-spherical-symmetric case, and of Reissner's solution, which was only an expansion of Schwarzschild's solution where the electrostatic field was also present. There were singularities in the original forms of both solutions. Einstein and Rosen removed these by removing the denominators from the solutions. The authors presented their results and thoughts about their new theory in the following words:

> If one solves the equations of the general theory of relativity for the static, spherically symmetric case, with or without the electrostatic field, one finds that singularities occur in the solutions. If one modifies the equations in an unessential manner so as to make them free from denominators, regular solutions can be obtained, provided one treats the physical space as consisting of two congruent sheets. The neutral, as well as the electrical particle *is a portion of space connecting the two sheets (bridge)* [emphasis—L.K.]. In the hypersurfaces of contact of the two sheets the determinant of the $g_{\mu\nu}$ vanishes.

> One might expect that processes in which several elementary particles take part correspond to regular solutions of the field equations with

several bridges between the two equivalent sheets corresponding to the physical space. Only by investigations of these solutions will one be able to determine the extent to which the theory accounts for the facts. For the present one cannot even know whether regular solutions with more than one bridge exist at all.

It appears that the most natural electrical particle in the theory is one without gravitating mass. One is therefore led, according to this theory, to consider the electron or proton as a two-bridge problem.

In favour of the theory one can say that it explains the atomistic character of matter as well as the circumstance that there exist no negative neutral masses, that it introduces no new variables other than the $g_{\mu\nu}$ and $\varphi_\mu$, and that in principle it can claim to be complete (or closed). On the other hand one does not see *a priori* whether the theory contains the quantum phenomena. Nevertheless one should not exclude *a priori* the possibility that the theory may contain them.

In any case here is a possibility, for a general relativistic theory of matter which is logically completely satisfying and which contains no new hypothetical elements.[369] **e40**

The last words of the passage quoted above indicate that Einstein was convinced he was going in the right direction. He maintained this conviction for a long time. In March of 1936, a year after the article was published, he popularised his and Rosen's new theory in the final part of the article entitled "Physics and Reality."[370] In an article entitled "Gravitational Equation versus the Problem of Motion," written in collaboration with Leopold Infeld and Banesh Hoffmann and published in 1938, almost three years later, Einstein quoted the article he had written in collaboration with Rosen as presenting views that were still valid.

## 4.7    History of ether continued in relativity theory

The book *The Evolution of Physics*,[371] of which Einstein was the co-author, relates the continuation of the history of the ether in the theory of relativity. The book was actually written, after thorough discussion about its contents with Einstein, by a young Polish physicist of Jewish origin, Leopold Infeld. The book was published in 1938.

---

[369] *Ibid.*, p. 77.

[370] A. Einstein, "Physik und Realität," *JFI*, 221 (1936), pp. 313-347.

[371] A. Einstein, L. Infeld, *Die Physik als Abenteuer der Erkenntnis* (Leiden: A. W. Sijthoff, 1938); English version: *The Evolution of Physics* (New York: Simon and Schuster, 1938).

Infeld met Einstein for the first time in Berlin in 1920, as a fifth-year student of physics at the Jagellonian University. He intended to complete his studies at Berlin University, where the lectures were delivered by such distinguished scientists as Planck, Laue and Einstein.[372] Because of German hostility towards Poles, he was unable to enrol in the university. Infeld asked Einstein to help him, and the senior scientist gave him a letter of recommendation addressed to Planck. Yet because of the anti-Semitic attitude of the university administration, he was not admitted.[373] Nevertheless, as the outstanding Polish physicist and historian of physics Bronisław Średniawa has shown,[374] Infeld managed to attend the lectures of Laue and Planck, and during his stay in Berlin wrote his doctoral dissertation "Light Waves in the Theory of Relativity."[375] While in Berlin, Infeld learned that Einstein had introduced the new concept of the ether, and consequently in his dissertation he wrote:

> Propagation of each action, whose background is the aether, can take place only along null geodesic lines in our four-dimensional continuum.[376] **e40a**

Infeld's dissertation was the first work in Polish on General Relativity.[377] More details concerning Infeld and the reception of the theory of relativity in Poland can be found in Średniawa's papers.[378]

The collaboration between Infeld and Einstein started in 1936 in the United States, where the two physicists met for the second time. After a

---

[372] L. Infeld, *Szkice z przeszłości* (Warszawa: Państwowy Instytut Wydawniczy, 1964), pp. 54-55.

[373] *Ibid.*, p. 63-102.

[374] B. Średniawa, "The reception of the Theory of Relativity in Poland," in: *Thomas F. Glick (ed.), The Comparative Reception of Relativity* (Dordrecht/Boston/Lancaster/Tokyo: D. Reidel Publishing Company, 1987), pp. 333-334.

[375] Infeld L., "Light Waves in the Theory of Relativity." *Prace Matem. Fiz.*, 32 (1922), pp. 33-84.

[376] *Ibid.*, p. 38.

[377] B. Średniawa, "The reception of the Theory of Relativity in Poland," in: Thomas F. Glick (ed.), *The Comparative Reception of Relativity* (Dordrecht / Boston/Lancaster/Tokyo: D. Reidel Publishing Company, 1987), pp. 327-350.

[378] B. Średniawa, "The Evolution of the Concept of Ether and the Early Develpment of Relativity at Cracow University," in: *Universitas Iagellonica Acta Scientiarum Litterarumque* MCLI, *Universitatis Iagellonicae Folia Physica, Fasciculus* XXXVII (Kraków, 1994), pp. 9-20; B. Średniawa, "Early Investigations in the Foundations of General Relativity," in: *Universitas Iagellonica Acta Scientiarum Litterarumque* MCLI, *Universitatis Iagellonicae Folia Physica, Fasciculus* XXXVII (Kraków, 1994), pp. 21-37; B. Średniawa, "Kontakty naukowe i współpraca polskich fizyków z Einsteinem," *Kwartalnik Historii Nauki i Techniki*, Vol. 41 (1996), No 1 p. 59-97.

year-long collaboration, which resulted, among other things, in the article mentioned in the previous section, Infeld ran into financial problems. He was not granted a scholarship for the academic year 1937-38. Einstein offered him assistance equal to the amount of the scholarship. Infeld did not accept the offer. He wanted to make a living himself, but it was not easy to get a job. He suggested that they should write a popular book together. Einstein accepted the proposal, and very willingly helped Infeld to write the book. This is how *The Evolution of Physics* (the German original was entitled *Physik als Abenteuer der Erkenntnis*) came to be written. The allowance Infeld received from the publishing house solved his financial problems.[379]

According to Peter Bergmann, who started a collaboration with Einstein at that time, Infeld discussed every sentence of *The Evolution of Physics* with Einstein.[380] Infeld later wrote:

> I was also aware that if the book was to have an historical value, I should remain in the shadows and allow Einstein to express his personal thoughts.[381] **e41**

In Bergmann's view, *The Evolution of Physics* contains a final excommunication of the old ether. After discussing the problem of the ether in the old sense of the term at great length (suggested even by the titles of some of the sections, *e.g.* "The Ether and the Mechanical View," "Field and Ether," "Ether and Motion") the book reads as follows:

> All our attempts to make ether real failed. It revealed neither its mechanical construction nor absolute motion. Nothing remained of all the properties of the ether except that for which it was invented, i.e., its ability to transmit electromagnetic waves. Our attempts to discover the properties of the ether led to difficulties and contradictions. After such bad experiences, this is the moment to forget the ether completely and to try never to mention its name.... The omission of a word from our vocabulary is, of course, no remedy. Our troubles are indeed much too profound to be solved in this way![382] **e42**

---

[379] L. Infeld, *Szkice z przeszłości* Warszawa: Państwowy Instytut Wydawniczy, 1964, p. 63-102.

[380] Information received from Peter Bergmann during the International Conference on History of General Relativity, Marseille, Luminy, 1988.

[381] L. Infeld, *op. cit.*, pp. 54-55.

[382] A. Einstein, L. Infeld, *Die Physik als Abenteuer der Erkenntnis* (Leiden: A. W. Sijthoff, 1949), p. 116. The quotation is taken from: *The Evolution of Physics* (New York: Simon and Schuster, 1961), pp. 175-176.

Although the ether in the old sense of the term has to be rejected, this does not mean that the history of the ether ended with the appearance of the theory of relativity.

> We may still use the word ether, but only to express the physical properties of space. This word ether has changed its meaning many times in the development of science. At the moment it no longer stands for a medium built up of particles. Its story, by no means finished, is continued by the relativity theory.[383] **e43**

The above passage reveals that Einstein still treated the space-time continuum as a new ether, for the same reasons as on the previous occasion, despite the fact that he rejected the unitary field theory with teleparallelism. In the new attempts to enrich its structure, the space-time continuum was understood as a total field endowed with real physical properties, energy included. Since it possessed energy, it also had mass, and this is why there was no quantitative difference between the total field and matter consisting of particles. The total field was also matter, though of a special kind, not consisting of particles in any sense. The term "ether" properly expressed the features of the space-time continuum understood as a distribution of energy momentum of a special kind.

## 4.8    Material nature of the space-time continuum

In their book *The Evolution of Physics*, in the subchapter entitled "Field and Matter," Einstein and Infeld emphasised that there could not exist any quantitative difference between field and matter:

> From the relativity theory we know that matter represents vast stores of energy and that energy represents matter. We cannot, in this way, distinguish qualitatively between mass and field, since the distinction between mass and energy is not a qualitative one. We could therefore say: Matter is where the concentration of energy is great, field where the concentration of energy is small. But if this is the case, then the

---

[383] *Ibid.*, pp. 99-100. The quotation is taken from the English version of this book: *The Evolution of Physics* (New York: Simon and Shuster, 1961), p. 153; however, I have made a correction on this point where it did not correspond to the German original. In the English version we find "some physical property of space" but in the German original we have "*physikalischen Eigenschaften des Raumes.*" Consequently, I have translated this as "physical properties of space." In my opinion the German original better expresses Einstein's idea of the new ether. Note that the German original and the English version were published simultaneously, in 1938, in Holland by A. W. Sijthoffs Witgenersmaatschappij, N. V. (German original), in the United States by Simon and Schuster, Inc., and in England by the Cambridge University Press (English version).

difference between matter and field is a quantitative rather than a qualitative one. There is no sense in regarding matter and field as two qualities quite different from each other. We cannot imagine a definite surface separating distinctly field and matter. [...] In the light of the equivalence of matter and energy the division in matter and field is something artificial and not well defined. [...] Matter is where the concentration of energy is high, field is where the concentration of energy is low. But if this is the case, the difference between matter and field is quantitative and not qualitative.[384] **e44**

## 4.9   New attempt to improve Kaluza's theory

Einstein collaboration with Bergmann resulted in a new attempt to improve Kaluza's five-dimensional theory. A description of the improvements can be found in the article "Generalisation of Kaluza's Theory of Electricity"[385] published in July of 1938. It differed from Kaluza's theory in one significant element: the authors assigned a certain physical reality to the fifth dimension, whereas in Kaluza's original theory the fifth dimension was introduced only formally, to generate the new components of the metric tensor that were to represent the electromagnetic field. Kaluza assumed dependence of field variables on four co-ordinates $x^1$, $x^2$, $x^3$, $x^4$, and not on the fifth co-ordinate $x^0$, when a suitable co-ordinate system was selected. Einstein and Bergmann, on the other hand, tried to show that it was possible to assign a certain physical sense to the fifth dimension, without causing any incoherence with the four-dimensional character of the physical continuum. In the new approach, it was assumed that physical space was closed with respect to the fifth co-ordinate. As a result, the fifth dimension ceased to be something purely formal and acquired a certain physical sense. According to the authors, this innovation brought about a significant simplification of the whole theory.

## 4.10   Einstein finally rejects Kaluza's theory

In 1941, Einstein's last work on Kaluza's unitary field theory, entitled "Five-dimensional Representation of Gravitation and

---

[384] *Ibid.*, p. 162-164. The quotation is taken from *The Evolution of Physics* (New York: Simon and Schuster, 1961), pp. 241-242.

[385] A. Einstein, P. Bergmann, "Generalisation of Kaluza's theory of electricity," *AoM*, 39 (1938), pp. 683-701.

Electricity,"[386] was published. Bergmann and Valentin Bargmann were co-authors, together with Einstein. The result of the study was negative. At the conclusion of the paper, the three authors stated that the formulas they had arrived at indicated that in Kaluza-Klein's theory it was very difficult to obtain an experimentally observable difference between the gravitational and electromagnetic interactions. The experimental evidence spoke against the theory.

## 4.11   The theory of bivector fields

In 1943, Einstein made another attempt to generalise the General Theory of Relativity. Two of his articles: "Bivector Fields I" and "Bivector Fields II," were published in that year.[387] Bargmann was the co-author of the first article. In the introduction, the authors stressed that all the attempts to generalise the General Theory of Relativity by introducing minor changes (for example, by replacing the four-dimensional space-time continuum with more-dimensional spaces) had been unsuccessful. As a result, if a new way to solve the unification problem was to be found, a more significant change of the essential features of the General Theory of Relativity had to be made.[388] The new theory of bivector fields entailed such a change. There are three essential elements of the original General Theory of Relativity:

(1) four-dimensional character of space-time continuum;
(2)   covariance of the field equations with respect to all continuous transformations;
(3)   existence of Riemann's metric (*i.e.*, symmetric tensor $g_{\mu\nu}$), which defines the structure of the physical continuum.

In the theory of bivector fields, the authors maintained the first two elements, but to describe the structure of space, they used a mathematical object which, though different from the third element of the General Theory of Relativity was, to a certain extent, similar to the symmetric tensor $g_{\mu\nu}$. This new mathematical object was called the basic symmetrical bivector, and denoted $g_{\mu\nu \atop \alpha\beta}$.

---

[386] A. Einstein, V. Bargmann, P. G. Bergmann, "Five-dimensional representation of gravitation and electricity," in: *Theodore von Karman Anniversary Volume* (Pasadena: California Inst. of Technology, 1941), pp. 212-225.

[387] A. Einstein, V. Bargmann, "Bivector Fields I," *AoM*, 45 (1944), pp. 1-14; A. Einstein, "Bivector Fields II," *AoM*, 45 (1944), pp. 15-23.

[388] A. Arodź, *op. cit.*, p. 307.

Many features of the theory of symmetrical bivector fields were similar to those of Riemann's theory with metrics. The essential difference, however, consisted in the fact that, unlike Riemann's $g_{\mu\nu}$, the basic variables of the $g_{\substack{\mu\nu \\ \alpha\beta}}$ field depended on the combination of the coordinates of two points. The authors were not sure, however, whether their new theory could give a suitable solution to the problem. They expressed their doubts in the introduction to the article.

The second article of the pair was Einstein's attempt to simplify the new theory by splitting the concept of relations into relations that were based exclusively on affine connection, and relations in which affine connection was limited by the hypotheses on field structure.

## 4.12   A new attempt to generalise General Relativity

Einstein was satisfied with the bivector theory for only a short time. In June of the following year (1945), his new article entitled "Generalisation of the Relativistic Theory of Gravitation"[389] was accepted for publication. It opened a series of studies[390] (some of them written in collaboration with Ernst Straus in the years 1945-1948) proposing the use of Hermite's metric tensor $g_{\mu\nu} = \overline{g}_{\mu\nu}$ for unification purposes.

In this theory the total field was described by tensor $g_{\mu\nu}$ with complex components. They fulfilled a symmetry condition that was a natural generalisation of the symmetry condition of the gravitational theory metric field. Einstein called this "Hermite's symmetry." The components of the complex metric tensor were continuous functions of co-ordinates $x_1, \dots x_4$. From $g_{\mu\nu} = \overline{g}_{\nu\mu}$, it followed that $g_{\mu\nu}$ was split into

$$g_{\mu\nu} = s_{\mu\nu} + i a_{\mu\nu}$$

where $s_{\mu\nu}$ and $a_{\mu\nu}$ fulfilled the conditions:

$$s_{\mu\nu} = s_{\nu\mu} \text{ and } a_{\mu\nu} = -a_{\nu\mu}.$$

For the transformation group of the proposed theory, $s_{\mu\nu}$ and $a_{\mu\nu}$ were independent tensors. Einstein was willing to interpret the

---

[389] A. Einstein, "Generalisation of the relativistic theory of gravitation I," *AoM*, 46, 1945, pp. 578-584.

[390] A. Einstein, E.G. Straus, "Generalisation of the relativistic theory of gravitation II," *AoM*, 47, 1946, pp. 731-741.

asymmetrical part of Hermite's tensor $g_{\mu\nu}$ as a representation of the electromagnetic field.[391] This time, however, he was more careful in drawing conclusions.

In the new attempt at unification, the total field appeared not as a unified invariant entity, but as two entities unified only formally. According to Einstein, this unification was of a limited significance.

We may add here that the term "total field" (*Gesamtfeld*, and, in the articles written in English, *entire field* or *total field*) now became dominant in Einstein's works. As a result, the previously preferred term (four-dimensional) "physical space" became less significant. The term "ether" never appeared in Einstein's work again. This does not mean, however, that Einstein stopped treating the total field as a special kind of matter. Just the opposite: the type of materiality Einstein assigned to the total field contributed to a certain change in his views on physical space. He strongly emphasised that empty space constituted not a *vacuum*, but a *plenum*. As we will see later, Einstein even came to believe that space did not have an autonomous existence, and was instead only a structural property of the field.

## 4.13 Asymmetric field—return to the 1925 idea

In 1950, Einstein came back to the asymmetric field theory with a real but asymmetric metric tensor ($g_{\mu\nu} \neq g_{\nu\mu}$). He improved and developed it down to his last days. The first work referring back to the 1925 idea was entitled "Generalised Theory of Gravitation." It was published as a supplement to Einstein's book *The Meaning of Relativity*,[392] which was a new edition of his four lectures on the theory of relativity delivered at Princeton University in May, 1922. The subsequent editions of this book (one in 1950, two in 1953 and one in 1955[393]) contained the same supplement with the improved theory of the asymmetric field.

Some of Einstein's works were written in collaboration with Bruria Kaufman.[394] The asymmetric field theory raised objections, which

---

[391] A. Einstein, "Generalisation... I," p. 583.

[392] A. Einstein, "Generalised Theory of Gravitation." As Appendix II to *The Meaning of Relativity* (Princeton: Princeton Univ. Press, 1950).

[393] A. Einstein, "Asymmetric field theory." As Appendix II to to *The Meaning of Relativity*, Fifth edition (Princeton: Princeton Univ. Press, 1950).

[394] A. Einstein, B. Kaufman, "Algebraic properties of the field in the relativistic theory of the asymmetric field," *AoM*, 59 (1954), pp. 230-244.

Einstein addressed in an article published in the *Physical Review*.[395] Some articles popularising the development of the 1925 theory were also published, the first of them in *Scientific American*[396] in 1950.

Einstein still tried to find a singularity-free solution that would allow him to develop a field theory of matter consisting of particles. Although he did not succeed in developing a field theory of matter within the asymmetric field theory, he was convinced that, as he advanced the theory, he was getting close to solving the problem of unification. In 1954, a year before his death, Einstein wrote:

> After long probing, now I believe that I have now found the most natural form for this generalisation [of General Relativity—L.K.].[397]
> **e45**

## 4.14   Changes in Einstein's views on physical space

When Einstein was developing his geometrical theory of unification (Riemann's geometry with teleparallelism), the concept of a real (four-dimensional) physical space appeared quite frequently in his works. At that time Einstein treated physical space as a primary entity. Physical space took over the functions of the ether (as it was essentially identical to the ether), incorporated time and even sought to incorporate elementary particles. Fields were treated as states of space, and particles as portions of space. In the final stage of his research, Einstein's total field—as we have mentioned before—became the *leitmotif* in all his works. Einstein tried to develop a field theory of the space-time continuum and matter. The total field became a primary entity for him. The physical space-time continuum was just a structural property of the total field, characterised by a special kind of materiality. The total field had a certain intensity, even in areas of so-called vacuum, and an energy density of its own. In short, there was no space without a field with its special materiality. Consequently, in an article written in 1950 and published in *Scientific American*, Einstein could write:

> According to general relativity, the concept of space detached from any physical content does not exist. The physical reality of space is

---

[395] A. Einstein, "A comment on a Criticism of Unified Field Theory," *PhR*, 89 (1953), p. 321.

[396] A. Einstein, "On the generalised theory of gravitation," *Scientific American*, 182 (1950), pp. 13-17.

[397] A. Einstein, "Relativity and the Problem of Space," in: A. Einstein, *Ideas and Opinions* (New York: Crown Publishers, Inc., Fifth Printing 1960), p. 376.

represented by a field whose components are continuous functions of four independent variables—the coordinates of space and time.[398] **e46**

Similar statements can be found in the 1954 article entitled "Relativity and the Problem of Space."

[...] space as opposed to "what fills space," which is dependent on the co-ordinates, has no separate existence [...]. There is no such thing as an empty space, i.e., a space without field. Space-time does not claim existence on its own, but only as a structural quality of the field.[399] **e47**

The 1955 "Supplement III" to the fifth edition of *The Meaning of Relativity* clearly stated:

[...] space has lost its independent physical existence, becoming only a property of the field.[400]

## 4.15 Did Einstein stop using the term "ether" after 1938?

The absence of the term "ether" in its new meaning in Einstein's works after 1938 (in which he said that the next stage of the unfinished history of the ether was found in the theory of relativity) was a fact. Only one hint of this new concept can be found there, in the 1954 article quoted above:

This rigid four-dimensional space of the special theory of relativity is to some extent a four dimensional analogue of H.A. Lorentz's rigid three-dimensional ether.[401] **e49**

In response to the question: Did Einstein stop using the term "ether" after 1938?, we note that in his strictly scientific or professional works, Einstein never actually used the term "ether" in the new sense of the word. Nor did he mention it in his lectures or the numerous interpreting works he published. He used only the term "total field." We cannot conclude, however, that he stopped interpreting space-time continuum models as models of the new ether, especially when we know that at that time he published new editions of two significant works on the relativistic ether, introducing a quite significant change to one of them. We can be sure of only one thing: in his final years he did not write any new texts on the relativistic ether. He was content to prepare new,

---

[398] A. Einstein, "On the generalised theory...," p. 15. A. Einstein, "Relativity and the Problem of Space," pp. 375-376.

[399] A. Einstein, "Relativity and the Problem of Space," pp. 375-376.

[400] A. Einstein, [As Appendix II to the fifth edition of *The Meaning of Relativity*].

[401] A. Einstein, "Relativity and the Problem of Space," p. 372.

revised editions of previously written works. We will discuss these new editions of older texts on the ether in the next section. At this juncture, we shall consider the psychological and objective reasons and arguments that may have caused Einstein to stop using the term "ether."

As we know, he was induced to use and publish his ether concept by Lorentz and Lenard. Einstein respected the first scientist very much, but had to defend himself against the attacks of the latter. Lorentz, whose scientific achievements Einstein greatly appreciated, had always advocated the resting ether, although he tried to modify his concept of the ether under the influence of the theory of relativity. In 1918, he proposed a model of the ether "permanently at rest." In his model, the ether was at rest with respect to any reference system, not just with respect to one distinguished system. Due to this feature, all reference systems were equivalent, and the principle of relativity was fully applicable to them.[402]

J. Illy[403] expressed the opinion that Einstein referred to his space-time continuum as "ether" out of respect for Lorentz: Einstein wanted to show his affinity with Lorentz's views and to honour the man, but when Lorentz died in 1928, Einstein gradually stopped using the term " ether."

A second psychological reason was pointed out by Don Howard.[404] When he was attacked by Lenard and other representatives of the so-called "German" school of physics, who were strong supporters of the ether, Einstein used the term "ether" to weaken their attacks, but he stopped using the term when he found that it was of no help to him.

We cannot say that the psychological reasons pointed out by Illy and Howard are totally false, because we do not know Einstein's real intentions. They may have played a certain role, both in incorporating the term "ether" into the theory of relativity, and in rejecting it again, especially after the term had become the banner in the fight against Einstein and his theory. But there may also have been more objective reasons behind his choice. We will attempt to describe them here.

When he noted that his time-space continuum had real physical properties described by the metric tensor $g_{\mu\nu}$, and perhaps recalling Drude's definition of the ether ("The ether is a space having physical properties."), in the favourable psychological climate created by correspondence with Lorentz and by the polemic with Lenard, Einstein recognised his new space-time continuum as a new relativistic ether, and,

---

[402] J. Illy, "Einstein Teaches Lorentz...," p. 272.
[403] *Ibid.*, p. 281.
[404] Prof. Don Howard conveyed this information to me.

for a limited period of time, became a strong supporter of the new concept, as can be seen mainly in his article "On the Ether" published in 1924. Moreover, at that time Einstein was not alone in supporting the relativistic ether. Other distinguished physicists, such as Weyl and Eddington, defended the concept in one way or another. This fact may have helped convince Einstein that it was advisable to call his space-time continuum an "ether."

Later on, however, when he noticed that the ether had been introduced into physics only because variable physical properties could not be assigned to space, and that the functions of the ether were easily taken over by the space-time of the General Theory of Relativity—even better, by the total field of his proposed theory of unification—Einstein may have come to the conclusion that it was not prudent to add unnecessary terms. Furthermore, he had to recognise a fact that he himself had brought about. In the 1940s and 1950s, influenced by the theory of relativity, physicists stopped using the term "ether." This may be the reason why Einstein ceased to use the term, especially when the terms "physical space" and "total field" were quite satisfactory. In his 1924 article, where, among other things, he said that in a complete field theory, all objects of physics would fall under the term "ether," Einstein wrote:

> [...] instead of speaking of an ether, one could equally well speak of physical qualities of space.[405] **e50**

This would appear to mean that Einstein did not pay much attention to names and terms. What mattered to him was the ideas. One need not use the term "ether," but one has to admit that the space-time continuum has real physical properties, with its own density or distribution of energy, if one treats it as a field. The idea of a special kind of materiality of the four-dimensional continuum, having space-time structural properties, was not weakened. Just the opposite: it was strengthened in Einstein's last works. Accordingly, the ether concept can be found in his last works, but under a different name. The prediction Einstein made in Leiden in 1920 did come to pass:

> The ether hypothesis must always play a part in the thinking of physicists, even if only a latent part.[406] **c51**

---

[405] A. Einstein, "Über den Äther," p. 85.

[406] A. Einstein, *Äther und Relativitätstheorie*, p. 4.

## 4.16   New editions of Einstein's works on the new ether

Einstein did agree to publication of two of his fundamental works on the new ether at the time when he was no longer using the term "the ether" in the relativistic meaning of the word. These two works were his lectures delivered in Leiden in 1920, and his article from the time when he thought that his geometrical unitary field theory was correct, *viz.*, the one published in 1934 in his book *Mein Weltbild.* It should be recalled, however, that although Einstein amended some of the new editions of his works, if the content was not in line with his current thinking, he did not correct the first of the two works for new editions, which were translated into many languages. This may mean that he did not cease interpreting the space-time continuum in the Special Theory of Relativity and General Theory of Relativity as the new ether. His lecture in Leiden gives this interpretation of both space-time continua. It is also confirmed by a hint at this interpretation, to be found in his 1954 article: there Einstein stressed that the four-dimensional time space of Special Theory of Relativity was analogous to Lorentz's three-dimensional ether.

The second of the two works noted above was extensively edited by Einstein in 1953. Einstein asked Carl Seelig, the editor (and author of numerous works on the scientist and his achievements) to remove the part that popularised his geometrical unitary field theory of 1928-1931 from the article.

> By request of Prof. Einstein, this had been left out, because "the theory there explained was abandoned by me a long time ago and it was replaced by the theory of the non-symmetric field, which is entirely satisfactory in a logical-formal sense."[407] **e52**

The revised article (1953) acquired a more general character. It was no longer linked to one unification scheme, and instead referred to all the new proposals. The statement that the space-time continuum and the ether were one and the same could now be referred to each attempt at unification, *i.e.*, each new mathematical representation of the space-time continuum as the total field of gravitational and electromagnetic interactions, including Einstein's theories of the asymmetric field.

---

[407] A. Einstein, "Letter to C. Seelig," 1953, quoted in: A. Einstein, *Mein Weltbild*, hrsg. von C. Seelig, mit Anmerkungen des Herausgebers (pp. 174-200) (Frankfurt/m-Berlin-Wien: Verlag Ullstein GmbH, 1983), p.199.

## 4.17 Did the idea of a relativistic ether survive?

April 18, 1955 will always be remembered as the day when Albert Einstein, one of the most distinguished physicists of all time, died. We have inherited the great wealth of his ideas. It is worth considering whether his concept of a relativistic ether has survived down to the present time. Our answer to this question must be positive, although the term "new ether" itself was—for a long time—almost forgotten in the community of scientists. However, the idea of a space-time continuum with a metric (or even richer) structure, with physical properties, actively participating in physical processes and having a certain density or distribution of energy (that is, materiality of a special kind) treated as a field, has survived; and it has even been improved along with the Special Theory of Relativity, the General Theory of Relativity, and in new attempts at unification as well. The belief that the so called "vacuum" has a structure of its own, that it actively participates in physical processes, and that it is not empty, has become a part of physics for good. Only the term "new ether" has almost disappeared, and many physicists even seem to have forgotten that Einstein ever used the term.

Nevertheless, there has recently been a revival of different ether theories, among them also relativistic ethers. Scientists have again acknowledged that Einstein was right in calling the space-time of relativity theory a "new ether." If we leaf through the proceedings of the international conferences on "Physical Interpretations of Relativity Theory,"[408] held every two years in London at Imperial College since 1988, and the Proceedings of other similar conferences, *e.g.*, "Frontiers of Fundamental Physics" in Olympia, Greece, 27-30 September 1993,[409] and "Open Questions in Relativistic Physics," Athens, Greece, 25-28 June 1997,[410] we find several papers concerned with both ether theories and relativistic ethers. (See also the book *The Philosophy of Vacuum* edited by Simon Saunders and Harvey R. Brown.[411]) Other renowned physicists

---

[408] M. C. Duffy, (ed.) *Physical Interpretations of Relativity Theory*, British Society for Philosophy of Science, London, Proceedings of the international conferences of 1988, 1990, 1992, 1994, 1996 and 1998.

[409] M. Barone and F. Selleri eds., *Frontiers of Fundamental Physics* (New York-London: Plenum Press, 1994).

[410] F. Selleri, ed., *Open Questions in Relativistic Physics* (Montreal: Apeiron, 1998).

[411] S. Saunders and H.R. Brown eds. *The Philosophy of Vacuum* (Oxford: Clarendon Press, 1991).

have also been advocates of an ether. For example, as M. Jammer has indicated,[412] John S. Bell stated the following in a BBC radio interview:

> Yes, the idea that there is an aether, and those Fitzgerald contractions and Larmor dilations occur, and that as result the instruments do not detect motion through the aether—that is a perfectly coherent point of view ...[413] **e52a**

There are two basic interpretations of relativity theory: the Einsteinian and the Lorentzian. In Einstein's interpretation the ether is ultra-referential. It can never be considered as a reference frame, because only this kind of ether (as present-day supporters of this interpretation insist) does not violate the principle of relativity. One may thus say that Einstein's ether has survived until today. But Lorentz's ether has also survived. We may even speak of a revival of Lorentz's theory of the ether (*e.g.*, in the works of S. J. Prokhovnik,[414] F. Selleri,[415] Jean-Pierre Vigier[416]). In the Lorentz interpretation, one reference frame is privileged, *i.e.*, the reference frame of the stationary ether.

In Selleri's view,[417] we must treat the cosmic background radiation (or in some of his papers, radiation coming from all directions from different stars) as a fundamental reference frame. For instance, if we

---

[412] M. Jammer, "John Stewart Bell and the Debate on Significance of his Contributions to the Foundations of Quantum Mechanics," in: *Bell's Theorem and the Foundations of Modern Physics*, eds. A. Van der Merwe, F. Selleri, G. Tarozzi (Singapore, New Jersey, London, Hong Kong: World Scientific, 1992), p. 5.

[413] P.C.W. Davies and J. R Brown, eds. *The Ghost in the Atom* (Cambridge: Cambridge University Press, 1986), pp. 49-50.

[414] S. J. Prokhovnik, *Light in Einstein's Universe* (Dordrecht: Reidel, 1985); see also: S. J. Prokhovnik, "A Cosmological Basis for Bell's View on Quantum and Relativistic Physics," in: *Bell's Theorem and the Foundation of Modern Physics*, eds. A. Van der Merwe, F. Selleri and G. Tarozzi (New Jersey, London: World Scientific, 1990), pp. 508-514.

[415] F. Selleri, "Space-time Transformations in Ether Theories," *Z. Naturforsch.* 46a, 1990, pp. 419-425., see also: F. Selleri, "Special Relativity as Limit of Ether Theories," in: M. C. Duffy, (ed.) *Physical Interpretations of Relativity Theory* (London: British Society for Philosophy of Science 1990), pp. 508-514; see also: F. Selleri, "Inertial Systems and the Transformations of Space and Time," *Physics Essays*, Vol. 8, no. 3 (1995), pp. 342-349; see also: F. Selleri, "Theories Equivalent to Special Relativity," in: *Frontiers of Fundamental Physics*, M. Barone and F. Selleri, eds. (New York-London: Plenum Press, 1994).

[416] J.-P. Vigier, "New non-zero photon mass interpretation of Sagnac effect as direct experimental justification of the Langevin paradox," *Physics Letters A*, 234 (1997), pp. 75-85. See p. 80.; see also *Physics Letters A* 175 (1993), p. 269.

[417] F. Selleri, *Found. Phys.* 26 (1996), p.641; see also F. Selleri, "Space-time Transformations in Ether Theories," *Z. Naturforsch.* 46a (1990), pp. 419-425; see also F. Selleri, "Special Relativity as Limit of Ether Theories," in: M.C. Duffy (ed.), *Physical Interpretations of Relativity Theory* (London: British Society for Philosophy of Science, 1990), pp. 508-514.

move at great velocity with respect to this reference frame, we, as living beings, would be killed by this radiation because of the Doppler effect. With increasing velocity, all bodies would feel a greater radiation pressure. Thus, according to Selleri, in the real world, the proper frame of this radiation is the most fundamental. All material bodies and all celestial bodies (stars, planets, galaxies) move at small velocities relative to the radiation. Experimental measurements of velocities with respect to the cosmic background radiation have been taken by R. A. Muller's group.[418] Our Earth moves at a velocity of *ca.* 400 km per second, and our galaxy at a velocity of about 600 km per second with respect to this radiation. Muller titled his paper, published in *Scientific American,* "The Cosmic Background Radiation and the New Derivation of the Ether."

In Selleri's theory, the Lorentz transformation used by Einstein and Lorentz is only a special case of a more general transformation that requires a stationary ether.[419]

According to Vigier,[420] there exists an absolute inertial frame $\Sigma_0$, as recently revived in the literature.

As a historian of physics I prefer not to discuss here, in a book devoted to Einstein's ether concept, which of the two interpretations of relativity theory (Einsteinian or Lorentzian), and which of the two concepts of the ether (relativistic or stationary) is closer to truth. I would only like to stress that both have survived until today, and both have their ardent supporters. I would also like to add that new experimental and theoretical research will decide in the future who is right in this controversial issue: the followers of Einstein or the followers of Lorentz.

---

[418] R.A. Muller, "La radiazione cosmica di fondo and la deriva dell'etere," in: *Relatività e cosmologia,* edited by Tulio Regge (Milano: Le Scienze S.p.A. editore, 1981), pp. 74-84.

[419] F. Selleri, "Space-time Transformations in Ether Theories," *Z. Naturforsch.* 46a (1990), pp. 419-425, see also: F. Selleri, "Special Relativity as Limit of Ether Theories," in: M.C. Duffy, (ed.) *Physical Interpretations of Relativity Theory* (London: British Society for Philosophy of Science, 1990), pp. 508- 514; see also: F. Selleri, "Inertial Systems and the Transformations of Space and Time," *Physics Essays,* Vol. 8, no. 3 (1995), pp. 342-349; see also: F. Selleri, "Theories Equivalent to Special Relativity," in: *Frontiers of Fundamental Physics,* M. Barone and F. Selleri, eds. (New York-London: Plenum Press, 1994).

[420] J.-P. Vigier, "New non-zero photon mass interpretation of Sagnac effect as direct experimental justification of the Langevin paradox," *Physics Letters A* 234 (1997), pp. 75-85; see p. 80. See also *Physics Letters A* 175 (1993), p. 269.

# Chapter 5

# PHYSICAL MEANING OF EINSTEIN'S
# RELATIVISTIC ETHER

While chapters 3 and 4 presented the history of Einstein's concept of the ether, the present chapter seeks to present all the different models of the new ether and their various versions, and to distinguish the most important features of the ether. This will allow us to grasp and better understand the physical meaning attributed to the ether by Einstein. Before we deal with the different models and their characteristics, however, we shall briefly present Einstein's ideas concerning the role of models in physics. We shall also compare Einstein's modelling of the space-time continuum with its representations in contemporary theoretical physics.

## 5.1   Einstein stresses the model-like nature of physical cognition

Now that we have studied Einstein's views of different methods of physical cognition and his research practice as exemplified in his work, we may dare to say that Einstein was a supporter of a methodological orientation which we could call "structural modellism."

In Einstein's conception, the physical world has a dual structure: the structure of the physical qualities that characterise it (their general characteristic is quantity, and hence physics can introduce the notion of measurable physical quantities), and a nomological structure, i.e., the structure of the laws that govern it. In other words, the latter is the structure of interrelations between physical quantities. The task of a theoretical physicist is to construct models of both structures.

In the letter of November 28, 1930 (*i.e.*, in the period when he was constructing one of the versions of the relativistic ether model, for which he used Riemann's geometry, complemented by the notion of teleparallelism), Einstein wrote to one of the most representative thinkers of the neo-positivist Vienna Circle, Moritz Schlick, criticising his ideas:

> Indeed, physics supplies relations between sense experiences, but only indirectly. For me its essence is by no means exhaustively characterised by this assertion. I put it to you bluntly: Physics is an attempt to construct conceptually a model of the real world as well as of its law-governed structure.[421] **fl**

As we can see, constructing models of both structures represented the core of physics for Einstein. Each model was an attempt to reconstruct the real structure of physical phenomena and the laws that governed them. In the excerpt quoted above we note quite clearly Einstein's point of view on reality and physical cognition. Paraphrasing his words we could say: the real world has its structure; so do the laws that govern it. The task of physics is to construct models of both structures in order to know them. In physical cognition we deal with modelling of structures, and the models constructed by theoretical physics are also structures, as they constitute the representation of real structures with the use of mathematical structures.

## 5.2   Einstein's space-time models and contemporary physics

To model the physical structure of the space-time continuum, Einstein used the mathematical apparatus he had acquired during his studies and through self-education. He received considerable help in his self-education from his friend the mathematician Marcel Grossmann. Einstein had introduced the space-time physics of the Special Theory of Relativity without using the tensor calculus or a four-dimensional geometric model; the first to do so was Hermann Minkowski. Einstein— as we know—was at first quite reserved about Minkowski's method, treating it as "superfluous erudition." He fully appreciated its value when he learnt, with help from Grossmann, the tensor calculus and multi-

---

[421] A. Einstein, "Letter to M. Schlick, 18/11/1930," in: Don Howard, "Realism and Conventionalism in Einstein's Philosophy of Science: The Einstein-Schlick Correspondence," *Philosophia Naturalis* 21 (1984), H. 2-4, p. 628.

dimensional metric geometry of Riemann. In the paper of 1913, which Einstein wrote together with Grossmann, he started to utilise the tensor calculus and the method of four-dimensional non-Euclidean geometry. Einstein never abandoned these mathematical tools, and constantly strove to perfect them.

In 1916, the physics of space and time became the physics of the new ether for Einstein. In his works on the new ether—since they were of a purely interpretative nature—we find practically no mathematical symbols and formulae. The only symbols that occasionally appear in these articles are the basic metric tensor $g_{\mu\nu}$ and basic square formulae, which define the space-time metrics of the Special Theory of Relativity and General Theory of Relativity:

$$ds^2 = dx_1^2 + dx_2^2 + dx_3^2 - d(ct)^2$$

or

$$ds^2 = \sum g_{\mu\nu} dx_\mu dx_\nu$$

*i.e.*, the recipe for measuring and calculating the space-time interval between two infinitely close events (the so called linear element). These symbols and formulae appeared in Einstein's works on the new ether, because, according to his interpretation, they described the real metrics of the space-time continuum and defined its real structure, which determines the behaviour of the test particles, together with measuring rods and clocks. These real features, ascribed to space and time by Einstein, were the reason why he started to recognise the existence of the new ether again.

After Einstein's death the terminology changed considerably, and a mathematical formalism of the Special Theory of Relativity and General Theory of Relativity has been developed, together with new attempts to create a broader Unified Field Theory (so called super-unification). The utilisation of mathematical symbols has also been perfected. We present a concrete example: Einstein did not use a separate symbol to denote the basic metric tensor and a separate one for its components, as he used $g_{\mu\nu}$ for both. Today—although there are still some who use Einstein's convention—the metric tensor is denoted by the symbol $g$, whereas its components are labelled $g_{\mu\nu}$. This fact is connected with the creation of differential geometry (Cartan) expressed in a language independent of co-ordinates.

Since the middle of the twentieth century, contemporary physics has used a geometrical structure called the differential manifold to model the space-time continuum of the Special Theory of Relativity and General Theory of Relativity, as well as all macroscopic physical theories. A discussion of the history of this structure and its basic features may be helpful.[422]

In order to arrive at the notion of a differential manifold we must start with the notion of arithmetic space. The basic arithmetic space is a one-dimensional space $R^1$ whose points represent all real numbers. In due time there were introduced arithmetic spaces $R^n$ of multi-dimensional character $(n = 1,2,...)$ in which every point $p$ was associated with $n$-numbers of $p = (p_1, p_2, ... p_n)$, The spaces $R^n$ are metrisable, *i.e.*, we can introduce metrics into them. Important values of $x_2 - x_1$, where $x_1$ and $x_2$ are elements of the space $R^n$, satisfy the axioms of metrics, because we can simply measure or calculate the distance $x_2 - x_1$. Arithmetic spaces $R^n$ with metrics are called Euclidean spaces $E^n$. The metric constitutes a superstructure in relation to the space $R^n$. The metrics as such, as we shall see in a moment, are also superstructures in relation to differential manifolds. The problem of introducing metrics will be discussed later, and for now we will examine arithmetic spaces. It should be noted that Cartesian orthogonal space co-ordinate axes are numeral axes, and that the analytical geometry of Descartes operates in arithmetic spaces. These spaces were also used for the differential equation. The notion of a differential manifold started to take shape in the nineteenth century, along with the notion of the function. In 1854 Riemann, in his famous inaugural lecture, made a few comments which later played an important role in shaping the notion of manifold. He pointed to the need for research into functional spaces. In 1880, Poincaré made a further step towards the theory of differential manifolds by using certain topological techniques for the theory of differential equations. The first precise definition of a two-dimensional smooth manifold was given in 1913 by Weyl.

Later the definitions of arbitrary-dimensional and even indefinitely-dimensional manifolds were introduced. We distinguish various classes of

---

[422] M. Heller, "The Manifold Model for Spacetime," *Acta Cosmologica*, 10 (1981), pp. 31-51; M. Heller, "Relativistic Model for Spacetime," *Acta Cosmologica*, 10 (1981), pp. 53-69; M. Heller, "Space-time Structures," *Acta Cosmologica*, 6 (1977), pp. 109-128; M. Heller, *Theoretyczne Podstawy kosmologii* (Warszawa: PWN, 1988; J. Foster, J. D. Nightingale, *A Short Course in General Relativity* (London: Longman, 1979).

manifold differentiation denoted by the symbol $C^k$. Here we are speaking about a more or less smooth space. The class of smoothness is selected according to the requirements of the physical situation to be modelled with the manifold. We usually assume that the manifold is of the class $C^\infty$, which means that it is differentiated an arbitrary number of times. In the theory of relativity $k = 2$ is usually sufficient.

A differential manifold is defined in the following way. Set $M$ constitutes a differential manifold if:

1) It can overlap with the sub-set family $U = \{U_i\}$;

2) At every $U_i \in U$ there is a definite local co-ordinate system;

3) Between any two local co-ordinate systems defined upon any two nonoverlapping sub-sets

$$U_1, U_2 \in U \quad (U_1 \cap U_2 = \varnothing)$$

there is a differentiable (of the required class of smoothness) transfer from one co-ordinate system to the other and *vice versa* (within the overlapping area). We generally consider all the "permissible" systems of co-ordinates even if the smaller quantity is sufficient to cover $M$.[423]

In order to model the space-time continuum we use orientative and coherent manifolds. Each coherent manifold allows for exactly two orientations. The orientation of the map $(U, \varphi)$ of a given manifold $M$ is defined by the coordinate system for $\varphi(U) \in R^n$. The set of maps is called an atlas. The manifold is called orientative if we have such an atlas where each of its two maps $(U_1, \varphi_1)$ and $(U_2, \varphi_2)$, whose domains intersect $(U_1 \cap U_2 \neq \varnothing)$, are conformably oriented. The manifold defined with such an atlas is called oriented, and we use four-dimensional orientative manifolds to model the space-time continuum.[424]

Until 1950 the process of getting to understand the structure of manifolds was very slow and painstaking. Finally, in the mid-century it became a basic mathematical tool for modelling the space-time continuum, not only for contemporary physical theories, but even those treated as historical (*e.g.*, modelling of the space-time continuum of

[423] M. Heller, "Time and Causality in General Relativity," *The Astronomy Quarterly*, Vol. 7 (1990), pp. 65-86 (*cf.* p. 70).

[424] M. Heller, *Teoretyczne podstawy*..., p. 23.

Aristotelian and of Galilean physics).[425] As we can now see, it was a step-by-step process; all macroscopic theories of contemporary and historical physics (provided they are reconstructed in the language of modern mathematics) assume that the arena on which the physical processes occur has the structure of a differential manifold. The basic elements of such a manifold (points) are called events. Those events are defined by four coordinates in the local maps, which refer to each other as local reference systems. In each map of this type one coordinate is treated as the coordinate of time, and the remaining three are treated as spatial coordinates. Point-like events, which are the basic elements of the manifold, constitute idealisations of true events. As idealisations, they have no extension in time or space, and the manifold model of the space-time continuum is very general. It can serve, as noted above, as the basic arena for physical events in all the macroscopic physical theories. This general nature of the model makes it "too poor" for the physics of a specific type (like classical mechanics, the Special Theory of Relativity, or the General Theory of Relativity) "to happen" within it. This model lacks a definition of the operations of measurement of spatial distances and time intervals, which are so important for any specific type of physics. It needs a recipe for measuring, *i.e.*, a so-called metric (Greek *metreo* = to measure). Only when we introduce metrics upon the four-dimensional differential manifold do space-time distance calculations become feasible. We may deal with different methods of introducing metrics upon the manifold, depending on the physical theory. The geometrical superstructures are in a sense the mathematical equivalents of physical measurements.

Here we are interested in the metrical structures we should provide for differential manifolds in order to work with the Special Theory of Relativity and General Theory of Relativity. In other words we need a differential manifold that meets specific requirements (*i.e.*, the requirement of coherence and orientation) and a square form specifying the linear element $ds^2$ which would define the metrics of the Special Theory of Relativity, and which would define the metrics of the General Theory of Relativity. At present, in order to differentiate square forms (and, consequently, metrics) we use a so-called index characterising a square form, which is denominated by the letter $I$ and the number $(n - I)I$, which characterises its signature. When the manifold meets the requirement of coherence, then the index and signature, due to the

---

[425] *Ibid.*, p. 16.

linearity of the metrics, are identical for all $x \in M$. In this case we simply talk about an index or a signature of the manifold. When $I = 0$ (or $I = n$) we are dealing with the Riemann metrical manifold, otherwise we talk about the semi-Riemann manifold. The semi-Riemann manifold with the index $I = 1$ or $I = n - 1$, $n \geq 2$ is called the Lorentz manifold (or pseudo-Riemann manifold). This is the metrical manifold of the General Theory of Relativity. It is characterised by the Lorentz metric (or pseudo-Riemann metric) and it is denoted by the letter $g$. The Lorentz manifold is briefly denoted by the pair of symbols $(M,g)$ where $M$ is a four-dimensional differential manifold, and $g$ is the Lorentz metric for $M$. To be precise, the Lorentz metric is not a proper metric (it does not meet all the requirements of a metric) but a pseudo-metric; however it is generally acceptable to talk about the "Lorentz metrics." It is the pseudo-metric of the General Theory of Relativity and the Special Theory of Relativity, primarily treated as the limiting case of General Theory of Relativity. In other words, the vector spaces $T_x(M)$ tangent to the Lorentz manifolds are the Minkowski manifolds. The metric of the Special Theory of Relativity is a specific case of the Lorentz metric; it becomes the Minkowski metric (actually a pseudo-metric as well). When treated separately, the space-time continuum of the Special Theory of Relativity is represented by the pair $(M,\eta)$, where $M$ is a four-dimensional differential manifold, and $\eta$ is the Minkowski metric. We now return to the Lorentz metric (*i.e.*, to the field of the metric tensor $g$). It is very rich, as it comprises a few other structures (*e.g.*, conformal and affine structures).

The success of the Special Theory of Relativity and General Theory of Relativity with the Lorentz metrical structure proved the effectiveness of the Lorentz manifolds in modelling space-time physics. Many important physical results can be derived from the Lorentz structure in an almost trivial way. Therefore the standard model of macroscopic space-time in contemporary physics is represented by a four-dimensional differential manifold with the Lorentz metrical structure $(M,g)$.

The four-dimensional manifold $(M,g)$ constitutes, however, only a "kinematic arena" for relativistic physics. We must introduce dynamic limitations resulting from Einstein's equations for the field of gravitation into the model $(M,g)$ which provide for the existence of matter, if we want to operate with the complete General Theory of Relativity. According to the General Theory of Relativity, the distribution of matter in the space-time continuum and its motion influence the structure of time and space which is modelled with a metrical manifold. Einstein's equations of the gravitational field, which make allowance for matter, describe the dynamics of the General Theory of Relativity. Therefore, we also need a model of the distribution and motion of matter, which would provide for mass, density of matter, the field of the energy-momentum vector, the electric charge, *etc.*

In the physical space-time continuum model in his Special Theory of Relativity and General Theory of Relativity, and in his attempts to formulate a unitary relativistic field theory, Einstein could not apply the tools and methods of the contemporary theory of differential manifolds and the structures we use with them, because he simply did not know them in the form in which they are taught and applied today. Nevertheless, his Special and General Theory of Relativity still managed to incorporate the concepts contemporary mathematics (and physics) emphasise and express in a new and more precise language. Translated into the language of contemporary mathematics, the Special Theory of Relativity and the General Theory of Relativity would not lose their validity. Quite the opposite: they would reveal their cognitive value in a more complete way. What is more, the General Theory of Relativity has made a great contribution to the development of contemporary geometrical methods.

In this context, a question may arise: which mathematical structure of contemporary theoretical physics represents the entity Einstein called "the new ether?"

During the Second International Conference on the History of General Theory of Relativity, which took place in August 1988 in Luminy, near Marseilles, after a lecture delivered by the author of this book in which he presented an outline of the history of Einstein's ether, some participants claimed that there was no such structure, whereas others stated that the use of the word "ether" by Einstein was something quite unfortunate, as it could lead to misunderstandings due to the different meaning physicists insist on ascribing to it. Still others believed that there was no need for more terminology if the expressions "space-time continuum" and "four-dimensional metrical manifold" were sufficient. André Mercier, a physicist and philosopher, expressed his opinion that Einstein was absolutely right when he used the term "ether" because it perfectly reflected the nature of relativistic space-time. The last to take the floor was Peter Bergmann, Einstein's collaborator in the years 1936-1944, who replied to the question asked above in the following way:

Oh, I try to add to this discussion that in the last decades of his life Einstein was concerned with unitary field theories of which he created a large number of models. So I think he was very conscious of the distinction between the differential manifold (though he did not use that term) and the structure you have to impose on the differential manifold (metric, affine or otherwise) and that he conceived of this structure, or set of structures, as potential carriers of physical distinctiveness and including the dynamics of physics.

Now, whether it is fortunate or unfortunate to use for the latter the term like ether? I think simply from the point of view of Einstein and his ideas that in the distinction between the differential manifold as such and the geometrical structures imposed on it we could, if we want, use the term ether for the latter.[426]

The author is certain that Bergmann was right when he claimed that the differential manifold as such, which is used to model space-time without imposing upon it such structures as metrics, *etc.* cannot be treated as a mathematical structure representing Einstein's relativistic ether.

Bergmann was right, because the four-dimensional differential manifold as such is a mathematical structure of too general a nature, and it cannot physically define distinctive features of the space-time continuum without imposing metrics and other structures upon it. It is too general, because it can serve as an arena or a background for any macroscopic physical theory (and even perhaps a microscopic one, because the debate over the status of the differential manifold in microphysics is ongoing). By the act of imposing metrics (*i.e.*, the recipe for measuring space and time intervals) and other structures upon it, the structure enriched in such a way turns into something that represents distinctive physical features of the real space-time continuum. In the case of Einstein's ether, the issue concerned imposing metrics and other structures which are presently used with the Special Theory of Relativity and General Theory of Relativity. In this context, we quote a passage from Einstein's lecture delivered at Leiden in 1920:

According to the general theory of relativity, space without ether is unthinkable; for in such space, not only would there be no propagation of light, but also no possibility of existence for standards

---

[426] P. Bergmann, "A comment (recorded on tape) on my lecture," L. Kostro, "Outline of the history of Einstein's relativistic ether conception," delivered at the International Conference on the History of General Relativity, Luminy, France 1988.

of space and time (measuring rods and clocks), nor therefore any space-time intervals in the physical sense.[427] **f3**

Although in the above quotation Einstein did not use the terminology of contemporary mathematics, the quotation tends to confirm Bergmann's statement. The author believes it would be more consistent with Einstein's ideas if we assumed that the new ether is represented in the contemporary mathematical formalism not by the very structures themselves (metrics and others) which are imposed on the four-dimensional differential manifold, but by all these elements together. According to this description, the ether of the relativity theory—provided we agree to introduce the term—is represented neither by the differential manifold itself, nor by the structures introduced upon it, but by the latter together with the differential manifold.

## 5.3.   Three models of Einstein's relativistic ether

In the interpretative works by Einstein on the new concept of the ether we can distinguish three basic models of the relativistic ether, and the third comes in several versions.

The first is the model of the Special Theory of Relativity. Einstein identified it with the flat space-time of the Special Theory of Relativity, which—according to the terminology used in his time, and often used today as well—has pseudo-Euclidean metrics. Since, within the flat space-time continuum of the Special Theory of Relativity, there exist systems of co-ordinates in which the components of the metric tensor $g_{\mu\nu}$ are constant and represented by the symbol $\eta_{\mu\nu}$:

$$g_{\mu\nu} = \eta_{\mu\nu} = \begin{vmatrix} 1 & 0 & 0 & 0 \\ 0 & 1 & 0 & 0 \\ 0 & 0 & 1 & 0 \\ 0 & 0 & 0 & -1 \end{vmatrix}$$

Einstein gives the $\eta_{\mu\nu}$ components as the mathematical tool for describing the basic metrical behaviour of the ether of the Special Theory of Relativity. Since in reference systems in which $g_{\mu\nu} = \eta_{\mu\nu}$, test particles behave according to the first principle of dynamics, *i.e.*, they are at rest or move along straight lines with constant velocity, Einstein called the ether

---

[427] A. Einstein, *Äther und Relativitätstheorie* ..., p. 15.

of the Special Theory of Relativity "the inertial ether,"[428] and he pointed out that it shared a feature in common with the "ether of Newtonian mechanics." Due to its flatness, the inertial ether is extendable to infinity. It is also rigid and absolute, *i.e.*, the presence of matter and its motion does not exert any influence on its structure.

> The four-dimensional space of the special theory of relativity is just as rigid and absolute as Newton's space.[429] **f4**

Using contemporary terminology and symbols, we can formulate this statement thus: the pair $(M, \eta)$, where $M$ is the four-dimensional differential manifold, and $\eta$ is the Minkowski metric on $M$, represents the model of the ether of Special Theory of Relativity (*i.e.*, the model of the space-time continuum of the Special Theory of Relativity).

The second model of the relativistic ether is the model of General Theory of Relativity. Einstein identified it with the space-time continuum of General Theory of Relativity and mentioned that the most important tool for describing it was the symmetrical metric tensor $g$ with the components $g_{\mu\nu}$, which are continuous functions of the coordinates of the arbitrarily introduced systems. In the model of the General Theory of Relativity, the ether has a pseudo-Riemannian metric, and it can therefore be finite as far as its expandability is concerned, although it remains unlimited (like the surface of a sphere, which has no limits, yet is finite). The ether of the General Theory of Relativity is neither rigid nor absolute. The presence of matter and its motion exerts an influence on its structure, which is variable in time, and therefore flexible. Einstein called this type of ether the "gravitational ether" (*Gravitationsäther*).[430]

Using contemporary terminology and symbols we could say that the ether of the General Theory of Relativity is represented by the differential manifold $M$, upon which is imposed a differentiable field of the metric tensor $g$, in other words the Lorentz metric (also called the pseudo-Riemannian metric); therefore we may label it with the pair of symbols $(M, g)$.

The third model of the relativistic ether arises from Einstein's attempts to formulate the Unitary Field Theory. The model has as many versions as Einstein made attempts to carry out the unification, *i.e.*, seven.

---

[428] A. Einstein, "Fundamental ideas and problems of the theory of relativity," in: *Nobel Lectures*, Published for the Nobel Foundation (Amsterdam-London-NewYork: Elsevier Publishing Company, 1967), pp. 483-492 (see. p. 488).

[429] A. Einstein, "Das Raum-, Äther-...," p. 238.

[430] A. Einstein, *Äther und Relativitätstheorie...*, p. 14.

Their common feature is the fact that the symmetrical metric tensor with the components $g_{\mu\nu} = g_{\nu\mu}$ no longer fully describes the structure of the relativistic ether. It is intended to constitute not the field of gravitation alone, but the total field with the unified gravitational and electromagnetic interactions. In each of these versions of the ether another mathematical entity is used for description. For example, in the asymmetric field version the entity is an asymmetrical metric tensor with sixteen components $g_{\mu\nu} \neq g_{\nu\mu}$; in the bivector field version, the mathematical entity called the basic symmetrical bivector $g_{\mu\nu \atop \alpha\beta}$; and in still

another version, the complex Hermitian metric tensor $g_{\mu\nu} = \overline{g}_{\mu\nu}$.

If we wished to express briefly the versions of the third model of the relativistic ether in the language of contemporary theoretical physics, we could say that those versions were attempts to impose mathematical structures richer than the field of the symmetrical metric tensor $g$ upon the differential manifold $M$.

## 5.4.   Essential attributes of Einstein's ether

In order to grasp the characteristics Einstein attributed to the new ether, we must first of all introduce a few concepts used by him in his interpretative works. We have in mind here the notions of "dynamic" versus "static image" of the world of events, "relational" versus "box" notions of space, "reference spaces" versus "physical space as such" (or in other words "space as a whole" or "physical space" written in its singular form).

In order to present the features of the relativistic ether, Einstein used two images, *i.e.*, a "dynamic" and "static" image of the world of events and of the motion of bodies and reference systems.[431] Naturally, he used these images for other purposes as well: for instance, by the year 1913 he had used the dynamic image alone in presenting the Special Theory of Relativity and in his attempts at formulating the General Theory of Relativity, although—after some initial reservations—he also started to appreciate the static image, introduced by Minkowski. After 1913 Einstein utilised chiefly the static image.

What do we mean when we speak of dynamic and static images? Motion can be described in two ways: in the first image, which Einstein and Infeld called "dynamic," motion is described as a sequence of events

---

[431] A. Einstein, L. Infeld, *Die Physik als Abenteuer der Erkenntnis* (Leiden: A. W. Sijthoff, 1994), p. 140.

in the three-dimensional space continuum. In this image, the position of bodies and reference systems may change in time. We do not mix time and space, although we may be fully aware of the links between time and space which were discovered by the theory of relativity. We may, however, present the very same motion in another way, using the image which Einstein and Infeld called "static." We obtain a static image when we consider the straight and curved world-lines (histories) of bodies or points in the four-dimensional world (*i.e.*, in the four-dimensional space-time continuum). Motion is no longer presented as a sequence of positions in the three-dimensional spatial continuum, but as something which exists in the space-time continuum.

> The world of events can be described dynamically by a picture changing in time and thrown onto the background of the three-dimensional space. But it can also be described by a static picture thrown onto the background of a four-dimensional space-time continuum. From the point of view of classical physics the two pictures, the dynamic and the static, are equivalent. But from the point of view of relativity theory the static picture is more convenient and more objective.
>
> Even in the relativity theory we can still use the dynamic picture if we prefer it. But we must remember that this division into time and space has no objective meaning since time is no longer "absolute." We shall still use the "dynamic" and not the "static" language in the following pages, bearing in mind its limitations.[432] **f5**

In his interpretative works presenting the new ether, Einstein used both languages, or images, although, due to the popularising nature of these works, he preferred the dynamic image because in layman's terms the dynamic image is more accessible.

The notions of "relational" space and "box" space[433] are related to the dynamic description of motion, which had been used until Minkowski introduced the static description. Here we must resolve the following problem: Does motion mean only the movement of something with respect to something else? And if so, when there are no elements that can move relative to one another, does no space exist? Or does

---

[432] *Ibid.*, p. 140. The quote is taken from the English version, *The Evolution of Physics* (New York: Simon and Schuster, 1961), p. 208.

[433] A.Einstein, "Relativity and the Problem...," p. 362; A. Einstein, "Foreword," in: M. Jammer, *Concepts of Space, The History of Theories of Space in Physics* (Cambridge, Mass.: Harvard University Press, 1954), pp. XIII-XIV; 3rd enlarged edition (New York: Dover Publications, 1993).

motion mean the movement of something within something infinitely extended, which is seen as a kind of box? And in this case, if there are no elements in motion, the space, nevertheless, does really exist. In the first case we are dealing with the relational notion of space, in the second— with the box notion. Here is how Einstein discussed the two notions of space (The quotes that follow are taken from Einstein's "Foreword" to the book *Concepts of Space. The History of Theories of Space in Physics* by the well-known historian of physics Max Jammer.[434]):

> Now as to the concept of space, it seems that this was preceded by the psychologically simpler concept of place. Place is first of all a (small) portion of the earth's surface identified by a name. The thing whose "place" is being specified is a "material object" or body. Simple analysis shows "place" also to be a group of material objects. Does the word "place" have a meaning independent of this one, or can one assign such a meaning to it? If one has to give a negative answer to this question, then one is led to the view that space (or place) is a sort of material object and nothing else. If the concept of space is formed and limited in this fashion, then to speak of empty space has no meaning [...].
>
> It is also possible, however, to think in a different way. Into a certain box we can place a definite number of grains of rice or of cherries, *etc.* It is here a question of property of the material object "box," which property must be considered "real" in the same sense as the box itself. One can call this property the "space" of the box. There may be other boxes which in this sense have an equally large "space." This concept "space" thus achieves a meaning which is freed from any connection with a particular material object. In this way by a natural extension of "box space" one can arrive at the concept of an independent (absolute) space, unlimited in extent, in which all material objects are contained. Then a material object not situated in space is simply inconceivable; on the other hand, in the framework of this concept formation it is quite conceivable that an empty space may exist.
>
> These two concepts of space may be contrasted as follows:
>
> (a) *space as positional quality of the world of material objects;*
> (b) *space as container of all material objects.*
>
> In case (a), space without a material object is inconceivable. In case (b), a material object can only be conceived as existing in space; space

---

[434] M. Jammer, *Concepts of Space. The History of Theories of Space in Physics* (Cambridge, Mass.: Harvard University Press, 1954).

then appears as a reality which in a certain sense is superior to the material world.[435] **f6**

The notion of "box space" leads to acceptance of the existence of a multitude of spaces moving with respect to one another.

When a smaller box *s* is situated, relatively at rest, inside the hollow space of a larger box *S*, then the hollow space of *s* is a part of the hollow space of *S*, and the same "space," which contains both of them, belongs to each of the boxes. When *s* is in motion with respect to *S*, however, the concept is less simple. One is then inclined to think that *s* encloses always the same space, but a variable part of the space *S*. It then becomes necessary to apportion to each box its particular space, not thought of as bounded, and to assume that these two spaces are in motion with respect to each other.

Before one has become aware of this complication, space appears as an unbounded medium or container in which material objects swim around. But it must now be remembered that there is an infinite number of spaces, which are in motion with respect to each other. The concept of space as something existing objectively and independent of things belongs to prescientific thought, but not so the idea of the existence of an infinite number of spaces in motion relatively to each other.[436] **f7**

The notions of "box space" and "relational space" are free creations of our imagination. They were reconciled by Descartes after they had been enriched, to some extent, by Galileo and Newton.

Both space concepts are free creations of the human imagination, means devised for easier comprehension of our sense experience. These schematic considerations concern the nature of space from the geometric and from the kinematic point of view, respectively. They are in a sense reconciled with each other by Descartes's introduction of the coordinate system, although this already presupposes the logically more daring space concept (*b*). The concept of space was enriched and complicated by Galileo and Newton, in that space must be introduced as the independent cause of the inertial behaviour of bodies if one wishes to give the classical principle of inertia (and therewith the classical law of motion) an exact meaning [...]. In contrast with Leibniz and Huygens, it was clear to Newton that the space concept (*a*) was not sufficient to serve as the foundation for the inertia principle and the law of motion.[437] **f8**

---

[435] A. Einstein, "Foreword," pp. XIII-XIV.
[436] A. Einstein, "Relativity and the Problem...," p. 362.
[437] A. Einstein, "Foreword," pp. XIII-XIV.

The two notions of space, (a) and (b), are connected—each in its specific way—with the notion of a rigid physical object. This is what they have in common. Moreover, space as a vessel (*i.e.*, the biggest box without walls) is understood as a rigid body penetrating through everything, which constitutes the absolute reference system.

> When considering the mutual relations of the location of bodies, the human mind finds it much simpler to relate the locations of all bodies to that of a single one [...]. This one body, which is everywhere and must be capable of being penetrated by all others in order to be in contact with all, is indeed not given to us by the senses, but we devise it as a fiction for convenience in thought.[438] **f9**

We are dealing here with a certain relation between the box notion and the relational notion of space. Each motion occurs within space, not outside it, but also with respect to space, *i.e.*, with respect to the biggest box without walls, *i.e.*, with respect to absolute space.

It was Einstein's contention that the relativity theory overcame the notion of absolute Newtonian space, understood as the absolute inertial system separating the notion of space from the notion of a rigid material object, and linking it to the notion of the field introduced in physics by Faraday and Maxwell. The theory of relativity utilises the field notion of space as a whole, which also incorporates both features, "relation" and "box," though in a different way.

On the one hand, there exists only relative motion of bodies with respect to other bodies, and of reference systems with respect to other reference systems. There is no sense in discussing motion with respect to space as a whole. There is no absolute motion, and space as a whole does not constitute the reference system. On the other hand, motion always takes place within space as a whole, never outside it. The next section discusses the field notion of space in greater detail.

In his interpretative works dealing with the issue of the new ether, Einstein often refers to reference systems as "reference spaces" (*Bezugsräume*), thus underlining their infinite number, composition of points and divisibility into parts. These "reference spaces" are contrasted with "space as such" (*der Raum als solcher*) or "space as a whole" (*das Raumganze*), or—most frequently—"physical space" (*physikalischer Raum*), the latter in its singular form, simultaneously stressing its uniqueness, being not composed of points, and indivisible into parts, together with the fact that it cannot in any way constitute a reference system. We must

---

[438] A. Einstein, "Raum, Äther...," p. 173.

be very cautious, however, because the expressions "space as such" and "space as a whole" carry two meanings in Einstein's works which were not clearly pointed out by Einstein, even if they are contextually evident. Both meanings were used by Einstein to present the evolution of the notion of space. In Einstein's conception, as we know, physical objects (particularly rigid or practically rigid bodies) played a decisive role in forming the notion of space; this was reflected also in geometry, as he claimed, where we have the idealisations of physical objects. The point, the straight line, the plane are such idealisations. Therefore, geometry, particularly the geometry of the ancient Greeks, referred to quasi-objects (the idealisations of physical objects) excluding "space as such" (or space as a whole), the notion with which pre-scientific cognition operated, and which specifically exists in the works of Descartes and in Newtonian physics. In this context, when the expressions "space as such" and "space as a whole" emerge, Einstein is using them in their first meaning.

> It is clear that the concept of space as a real thing already existed in the pre-scientific conceptual world. Euclid's mathematics, however, knew nothing of this concept as such; it confined itself to the concepts of the object, and the spatial relations between objects. Point, plane, straight line, segment are solid objects idealised. All spatial relations are reduced to those of contact (the intersection of straight lines and planes, points lying on straight lines, *etc.*). Space as a continuum does not figure in the conceptual system at all. This concept was first introduced by Descartes, when he described the point-in-space by its coordinates. Here for the first time geometrical figures appear, in a way, as parts of infinite space, which is conceived as a three-dimensional continuum (...)

> In so far as geometry is conceived as the science of laws governing the mutual spatial relations of practically rigid bodies, it is to be regarded as the oldest branch of physics. This science was able, as I have already observed, to get along without the concept of space as such, the ideal corporeal forms—point, straight line, place, segment—being sufficient for its needs. On the other hand, space as a whole, as conceived by Descartes, was absolutely necessary to Newtonian physics. For dynamics cannot manage with the concepts of the mass point and the (temporally variable) distance between mass points alone. In Newton's equations of motion, the concept of acceleration plays a fundamental part, which cannot be defined by the temporally variable intervals between points alone. Newton's acceleration is only conceivable or definable in relation to space as a whole.[439] **f10**

---

[439] A. Einstein, "Das Raum-, Äther-...," pp. 233-234.

In the above quotation from Einstein's text devoted to the issue of space, the ether, the field, written in 1934, "space as such" is understood as that infinitely rigid, three-dimensional continuum which was used by Descartes to describe the position of points by introducing rectangular systems of co-ordinates. Newtonian physics attributed the character of something real to it, treating it as an absolutely rigid reference system interpreted as a complex of points at rest with respect to one another.

Einstein did not contrast this meaning of space as such, or of space as a whole, with reference spaces, because both the Cartesian continuum and the Newtonian absolute space were at the same time reference spaces and geometrical complexes composed of points, divisible into parts whose motion could be traced in time. Einstein avoided identifying his ether with space understood in this way, when (in the same article quoted above) he used the term "physical space" in the singular:

> Physical space and the ether are only different expressions for one and the same thing...[440] **f11**

To identify this space with the ether would violate the relativity principle, and the ether would become a preferred reference system. The prototype of the notion of space used by Descartes and Newton was the rigid extended body, understood as a complex of material points and divisible into elements. In Einstein's opinion, the old ether was frequently understood in this way.

The field notion introduced by Faraday and Maxwell was the next step in the development of the notion of space as such (or space as a whole). Initially, the field was imagined as something within matter, and thus a special material medium, the ether, was introduced to serve as carrier. Gradually, however, the field notion was freeing itself from the notion of a mechanical carrier.

> The emancipation of the field concept from the assumption of its association with a mechanical carrier finds a place among the psychologically most interesting events in the development of physical thought.[441] **f12**

In the theory of relativity, the field free from any mechanical carrier became the prototype of the notion of physical space as such, and expressed its physical meaning.

---

[440] *Ibid.*, p. 237.
[441] A. Einstein, "Relativity and the Problem...," p. 368.

According to general relativity, the concept of space detached from any physical content does not exist. The physical reality of space is represented by a field whose components are continuous functions of four independent variables—the coordinates of space and time. It is just this particular kind of dependence that expresses the spatial character of physical reality.[442] **f13**

As we can see, in the theory of relativity the field became the prototype of physical space, and its essence. It also contained the incentive to present all of reality within the field category. The victory over the notion of absolute space, understood as the basic inertial system, was won thanks to the rejection of the notion of the rigid physical object as a prototype of the notion of space, and replacing it with a new prototype—the field notion.

The victory over the concept of absolute space or over that of the inertial system became possible only because the concept of the material object was gradually replaced as the fundamental concept of physics by that of the field. Under the influence of the ideas of Faraday and Maxwell the notion developed that the whole of physical reality could perhaps be represented as a field whose components depend on four space-time parameters. If the laws of this field are in general covariant, that is, are not dependent on a particular choice of coordinate system, then the introduction of an independent (absolute) space is no longer necessary. That which constitutes the spatial character of reality is then simply the four-dimensionality of the field. There is then no "empty" space, that is, there is no space without a field.[443] **f14**

This is how the field notion of space, closely linked to time, was created. Physical space understood in this way did not constitute a reference system because it did not consist of points or parts whose history could be traced in time. Einstein contrasted "reference spaces," which constituted reference systems composed of points and parts, with "space as a whole" or "as such" in the other meaning of the word, *i.e.,* the meaning discussed above.

It should be noted that rejection of the rigid physical object as a prototype of the notion of physical space is discussed in Einstein's letter to Lorentz, in which he introduced his new notion of the ether for the first time:

---

[442] A. Einstein, "On the Generalised Theory of Gravitation," p. 348.

[443] A. Einstein, "Foreword," pp. XIII=XIV.

This new ether theory, however, no longer violates the principle of relativity, because the state of this $g_{\mu\nu}$ = ether would not be that of a rigid body in an independent state of motion. Every state of motion, instead, would be a function of position, defined by material processes.[444] **f15**

Here, Einstein rejects the old ether, understood as a rigid body whose motion relative to the Earth had been discussed, and treats the new ether as the field, which defined (by its own structure at a given place influenced by material processes) the state of the motion of test particles introduced into it.

S. Saunders and H. R. Brown are, therefore, correct when they call Einstein's new ether "the classical field-ether" and distinguish it from "the quantum field-ether"[445] because Einstein's ether theory is a classical, *i.e.*, pre-quantum theory.

By using the dynamical description of the world of events, Einstein identified his new ether with physical space understood as a field. It should be pointed out that whenever we use the expressions "space as such" and "space as a whole," we are using them in this very meaning (*i.e.*, the second meaning, not the first). Identifying his new ether with physical space as such, Einstein simultaneously stressed the fact that no state of motion could be attributed to it, because it did not consist of points or parts whose history could be traced in time.

> It does remain allowed, as always, to introduce a medium filling all space [...]. But it is not allowed to attribute to this medium a state of motion at each point, by analogy to ponderable matter. This ether cannot be conceived as consisting of particles that can be individually tracked in time [...] Since in the new theory, metric properties can no longer be separated from "truly" physical ones, the concepts of "space" and "ether" merge together.[446] **f16**

> Physical space and the ether are only different terms for the same thing: fields are physical states of space. If no particular state of motion can be ascribed to the ether, there does not seem to be any ground for introducing it as a substance of a special sort alongside space.[447] **f17**

---

[444] A. Einstein, "Letter to H. A. Lorentz," 17/6/1916, *EA* 16-453.

[445] S. Saunders and H. R. Brown, "Reflections on Ether," in: S. Saunders and H. R. Brown, eds., *The Philosophy of Vacuum* (Oxford: Clarendon Press, 1991), p. 29.

[446] A. Einstein, (*Morgan Manuscript*) *EA* 20-70.

[447] A. Einstein, "Das Raum-, Äther-...," p. 237.

The above-quoted texts clearly show that Einstein identified the new ether with space; it must be noted, however, that he did not identify it with any reference space. Identifying the relativistic ether with a reference space would have meant distinguishing one of them, thus violating the relativity principle. In Einstein's presentation of the Theory of Relativity, the distinction between "reference spaces" and "physical space as such" is clear and evident. It was not so in classical physics: "absolute space," which was linked or identified with the old ether, also constituted a reference space; it was understood, however, as a privileged absolute reference system. In the theory of relativity, physical space as such does not constitute a reference system; if physical space as such were the reference system, Einstein's ether would be the stationary ether like the ether of Lorentz, and we could speak of motion with respect to the ether. The relativistic ether, however, is not a stationary ether, and hence we cannot speak of motion with respect to it.

The concept of motion in the theory of relativity, including also the particular case of rest, can be applied only to all reference spaces, because only they are capable of moving with respect to one another, changing their mutual position in time. No state of motion, no rest, can be attributed to space as such. The concepts of motion and rest are entirely inapplicable here.

Einstein often stressed that physical space as such was connected with time as such. Therefore, completing Einstein's idea, we could say that his ether is related to time, which—it must be understood—is not composed of instants of time and time intervals. Only proper times connected with reference spaces are composed of instants of time and time intervals.

In the dynamic picture of the world of events, the point and space interval, and the instant of time and time interval play decisive roles. In the static description, however, the basic role is played by the "event" and by the space-time interval between events. In this description, the notion of world-line corresponds to the notion of the point, and the notion of momentary space to the notion of instants of time. The point in a given reference space is the place of co-local events, *i.e.*, the set of events following one another at the same place in the reference space under consideration. This set represents the line called world-line (history) in the static description. A moment of the proper time of a reference space is the set of all simultaneous events in the reference space under consideration. This set is represented by the three-dimensional section of the four-dimensional world, which is called a momentary space in the static description.

When he used the static image of the relativistic ether, Einstein identified it with the space-time continuum written in the singular, and he emphasised that the ether was not composed of world-lines. What follows is an excerpt from his lecture in Leiden in which he briefly contrasted the dynamic description with the static description:

> Extended physical objects can be imagined to which the idea of motion cannot be applied. They are not to be thought of as consisting of particles whose course can be followed out separately through time. In Minkowski's idiom this is expressed as follows: Not every extended entity in the four-dimensional world can be regarded as composed of world-lines.[448] **f18**

Thus, the ether which Einstein identified with space-time (written in the singular form) is not composed of world-lines and momentary spaces. Only reference space-times are composed of world-lines and momentary spaces.

Einstein laid great stress on the metrical structure of his ether, and therefore in his works on the ether the symbols of the metric tensor and its components appeared quite often. Einstein related its specific activity to the metrical structure of his ether, because it determined the behaviour of test particles, measuring rods, and clocks. In Einstein's conception of the ether we are dealing with the gradual activation of physical space, closely linked to time. Space and time used to be imagined as something passive and constant, similar to the indifferent vessel in which material physical processes occur. According to this image, space and time do not exert any active influence on what is happening in matter. In Einstein's conception of the ether, space and time ceased to be the passive and unchangeable arena where physical events are played out. The activation of time and space in Einstein's concept of the ether is three-staged, and at each stage the activity of space and time is represented as the activity of a specific kind of field.

We find the first stage of activation in the Special Theory of Relativity, *i.e.,* in the model of the ether of the Special Theory of Relativity. In this model, the relativistic ether actively determines the inertial behaviour of the particles within it, and thus constitutes the inertial field defined by the metric tensor $\eta$, which in a certain class of co-ordinate systems has constant components $\eta_{\mu\nu}$.

---

[448] A. Einstein, *Äther und Relativitätstheorie...*, p. 10.

We encounter the second stage of space and time activation in the General Theory of Relativity model of the ether. What this means is that the ether of the General Theory of Relativity determines both the inertial and the gravitational behaviour of particles. It constitutes the active inertial-gravitational field described by the symmetrical metric tensor $g$ with its components $g_{\mu\nu} = g_{\nu\mu}$, which are interpreted as gravitational potentials. The model of the General Theory of Relativity accomplishes a unification of inertia with gravitation. Einstein, as we know, was prepared to assign to the ether of General Theory of Relativity the active capability of producing elementary particles if he succeeded in finding the singularity-free solution of the equations of the gravitational field. For a brief period of time (1935-1937), Einstein was convinced that he, together with Rosen, had found this solution.

We are dealing with further attempts to activate space and time in all the attempts to formulate the Unitary Field Theory, *i.e.*, in all the versions of the third model of the ether. In these versions, the relativistic ether determines the inertial-gravitational behaviour of material particles, and also serves as an active medium in electromagnetic interactions, since it is understood as the total field for both kinds of interaction. In this third stage of activating time and space, Einstein made further attempts to find appropriate solutions, this time without singularities of the total field equations, which would represent the activity of the ether (identified with the total field) in producing elementary particles.

It must remembered that in the Special Theory of Relativity model of the ether, the activity of space and time is one-sided only. The space-time continuum of the Special Theory exerts an influence on physical processes, but they, in turn, exert no influence upon it. The situation in the model of the General Theory of Relativity, and in the different versions of the third model, is different.

Summing up this Section from the point of view of the contemporary mode of description, we might say that Einstein identified the activity of the new ether:

1) first, with the activity of the inertial field produced when the field of the $\eta$ tensor (*i.e.*, the Minkowski metric) is imposed upon the differential manifold $M$;

2) then, with the activity of the gravitational field produced when the field of the $g$ tensor (*i.e.*, Lorentz metric) is imposed upon the differential manifold $M$;

3) finally, with the activity of the total field that would be produced if Einstein had succeeded in his (sensible from the physical point of view) attempt to introduce the field of the asymmetric tensor $g_{\mu\nu} \neq g_{\nu\mu}$. Einstein sought to do this, first in 1925, and then in the final years of his life.

In Einstein's concept of the ether there also occurs a gradual materialisation of physical space-time. We must repeat here, however, that when we talk about materialisation of the space-time continuum, we mean ascribing it a specific type of materiality, very different from the materiality of the substances we encounter in physics, to which we refer when we use the word "matter."

At first glance, the ether of the Special Theory of Relativity seems deprived of any features of materiality. If we remove the matter composed of particles and electromagnetic fields from the space-time continuum of the Special Theory of Relativity, the continuum would appear absolutely empty. Nevertheless, it was the Special Theory of Relativity that supplied Einstein with a basic argument for attributing materiality to the space-time continuum. One of the premises of this argument is the principle of equivalence of mass and energy, which appears to be one of the major achievements of the Special Theory of Relativity. According to relativity theory, space-time is a field which has energy, and thus it also has mass, and therefore also a specific type of materiality. In light of the General Theory of Relativity, the space-time continuum of the Special Theory of Relativity constitutes an inertial field which is a specific, extreme case of the field of gravitation. The gravitational field within the so-called Galilean zones and in free-falling elevators is transformed into the local field of inertia. The field of inertia in non-inertial systems, *e.g.* in vehicles moving with accelerated motion, or on rotating platforms, is transformed into the local gravitational field.

> A space of the type (1) [of the special relativistic type—L.K.], judged from the standpoint of the general theory of relativity, is not a space without field, but a special case of the $g_{\mu\nu}$ field.[449] **f19**

Given the unification of inertia and gravitation accomplished by the General Theory of Relativity, we realise that the space-time continuum of the Special Theory of Relativity has the nature of a field, *i.e.*, it exhibits those features of materiality which Einstein attributed to fields because they possessed energy.

---

[449] A. Einstein, "Relativity and the Problem...," p. 375.

As we know, in the General Theory of Relativity, space-time constitutes a field of inertia and gravitation, which is characterised by a specific distribution of energy, and thus shares the materiality that is typical of fields. Since the space-time of the General Theory of Relativity does not cover electromagnetic fields with its structure, the ether of General Relativity is solely a gravitational ether. Electromagnetic fields have materiality of their own, complementary to the materiality of the ether of General Theory of Relativity, although the presence of these fields, due to their materiality, influences the structure of the gravitational ether. The presence of matter exerts an influence on the structure of the gravitational ether; the presence of matter exerts an influence on the structure of the space-time continuum of the General Theory of Relativity. Attempts to create a field theory of matter within the General Theory of Relativity led to an even greater stress on the materiality of the space-time of General Theory of Relativity, because within those attempts material particles were understood as "special states of space" or "portions of space." The materiality of the continuum possessing the properties of space-time was to find its highest expression in Einstein's attempts to formulate the Unitary Field Theory, since this was understood as a total field which covered the materiality of the field of gravitation as well as electromagnetic fields. Attempts to create a field theory of matter, where particles were created within the continuum understood as a total field, aimed at incorporating the entire physical reality into the concept of this material total field. These attempts in particular disclosed the qualitative identity of field and matter, which was often pointed out by Einstein.

The specific materiality of Einstein's gravitational field is described quite well by Roger Penrose in his paper "The Mass of the Classical Vacuum."[450] Penrose's presentation can be summarised in the following way. The energy density of the gravitational field in General Relativity is not measured in the standard way, which would be by the energy momentum tensor $T_{ab}$ appearing on the right side in the Einstein equation

$$R_{ab} - \tfrac{1}{2}Rg_{ab} = -8\pi GT_{ab}$$

(where $G$ is Newton's gravitational constant, $R_{ab}$ is the Ricci tensor, $g_{ab}$ the metric tensor, and $R$ the scalar of curvature). If in the vacuum (*i.e.*, in Einstein's gravitational field) there are many continuous media present

---

[450] R. Penrose, "The Mass of the Classical Vacuum," in: S. Saunders and H. R. Brown, eds., *The Philosophy of Vacuum* (Oxford: Clarendon Press, 1991), pp. 21-26.

(*e.g.*, electromagnetic fields, quantum field descriptions of particles) then we have an energy density (per unit volume), and hence a corresponding tensor, for each one. We add all these tensors together to obtain the *energy momentum tensor* $T_{ab}$ of the system. This tensor is supposed to describe the entire non-gravitational energy. In the absence of all physical fields except gravity, the energy momentum tensor $T_{ab} = 0$ and therefore also $R_{ab} = 0$; but it does not mean that there is no gravitational field present. The gravitational field existing in such conditions, *i.e.* the gravitational tidal distortion is described by the full *Riemann curvature tensor* $R_{abcd}$, which has a total of 20 components. The Ricci tensor has just ten components, and the remaining ten together form another tensor called the *Weyl tensor*

$$C_{abcd}$$

In the vacuum, such as inside a pure gravitational wave, the Weyl tensor still survives, and it can be considered as describing the free gravitational field. We must be aware, however, that the Weyl tensor is not an appropriate means for directly describing gravitational energy.

"Nevertheless," writes Penrose,

> ...gravity does contribute to the total mass-energy of a physical system. The simplest way of seeing this is to consider two masses. When they are far apart, the total energy of the system is somewhat greater than when they are close together, owing to the Newtonian potential energy contribution. Thus, by $E = mc^2$ they must have a slightly greater mass when they are far apart than when they are close together. The difference would have to come from the gravitational field in the space between the masses. But this cannot arise as an integral of the energy density locally defined in $T_{ab}$ because that energy density is zero outside the masses. Also, as mentioned above, (pure) gravitational waves carry energy, yet the energy density throughout the waves is every where zero.

These problems are related to the fact that the "conservation law"

$$\nabla^a T_{ab} = 0$$

that is enjoyed by the energy-momentum tensor is a "covariant" one ($\nabla^a$ denoting covariant derivative), and does not give rise to the integral conservation law that one would like, namely one asserting that the total energy of a physical system is actually constant. In a well-known attempt to resolve these issues, Einstein introduced a quantity $\Im_{ab}$, referred to as the energy momentum pseudo-tensor which was intended to take the energy of the gravitational field into account. This did give rise to an integral conservation law, but it suffered from the very serious drawback of depending heavily on the particular system of coordinates that happened to have been chosen for the problem at hand. The components of $\Im_{ab}$ therefore had no local physical meaning (a difficulty that was already appreciated by Einstein), and one certainly cannot take this pseudo-tensor description as providing the "true" measure of mass-energy distribution in the gravitational field.

One might take the view, nevertheless, that somehow the curvature of space-time-as measured by the surviving Weyl components of the curvature tensor-can still represent the "stuff" of gravitational waves. But, as is indicated by the above arguments, gravitational energy is non-local, which is to say that one cannot determine what the measure of this energy is by merely examining the curvature of space-time in limited regions. The energy-and therefore the mass-of a gravitational field is a slippery eel indeed, and refuses to be pinned down in any clear location. Nevertheless, it must be taken seriously. It is certainly there, and has to be taken into account in order that the concept of mass can be conserved overall.[451]

## 5.5    "Physical space," "ether," "field": are they synonymous?

The great importance of the relativity theory lies in the fact that it brought about the unification of several notions. What had seemed to be absolutely different turned out to be closely related, if not identical. Therefore, many notions that had seemed to signify various things having nothing in common with one another, became synonyms, or began to function as a single term comprising two names that were previously separate. Thus, new expressions were coined, such as "space-time" and "energy-momentum four-vector." The process of unifying the physical quantities of mass and energy culminated in the principle of equivalence of those two quantities.

A further consequence of the (special) theory of relativity is the connection between mass and energy. Mass is energy and energy has mass. The two conservation laws of mass and energy are combined by the relativity theory into one, the conservation law of mass-energy.[452]
**f20**

[451] R. Penrose, "The Mass of the Classical Vacuum," in: S. Saunders and H. R. Brown, eds., *The Philosophy of Vacuum* (Oxford: Clarendon Press, 1991), pp.24-25.

[452] A. Einstein, L. Infeld, *op. cit.*, (Leiden 1949) p. 132. The quote is taken from *The Evolution of Physics* (New York: Simon and Schuster, 1961), p. 197-198.

The thrust toward unification is also evident in Einstein's concept of the ether, and so we might ask at the end of this final chapter to what extent the expressions "physical space," "ether," and "field" denote one and the same thing.

In the ether models of the Special Theory of Relativity and of the General Theory of Relativity, and in the geometric Unitary Field Theory of 1928-1931 the terms "physical space," "ether," and "field" are basically synonyms. In the model of the Special Theory of Relativity, however, Einstein meant "the field of inertia," while in the General Theory of Relativity model it was "inertial-gravitational field," and in the above-mentioned Unitary Field Theory and others, "total field" was the meaning. It should be noted, however, that the expressions we are discussing here were used by Einstein in different periods with varying frequency, and they were attributed various ontological states of the reality they signified.

In the years 1918-1926 Einstein used the terms "physical space" and "ether" interchangeably, though he preferred to use the latter. He expressed the identity of the gravitational field with the ether, among other things, by referring to the latter as *the gravitational ether*. During this period, the new ether, in a sense, enjoyed a distinguished ontological status. Einstein planned to cover all the objects of physics with the notion of the ether.

In the years 1927-1934, the expression "physical space" became predominant, and the term "ether" fell into relative disuse, although Einstein still used the words interchangeably, as can be proved by a statement from that period, quoted earlier:

> Physical space and the ether are different terms for the same thing; fields are physical states of space.[453] **f21**

The ontological status of fields in relation to space was of secondary character. Fields were treated as states of space, although the physical content of space was exhausted in the physical meaning of the gravitational and electromagnetic field, particularly when, in 1931, Einstein introduced the expression "total field." Einstein's use of the word "ether" was rooted in his recognition that physical space absorbed the ether, thus taking over its functions. Therefore, during that period physical space gained privileged ontological status, and became "the sole carrier of reality."[454]

---

[453] A. Einstein, "Das Raum-, Äther-...," p. 237.
[454] A. Einstein, "Raum, Äther...," p. 180.

In the years 1935-1955, the expression "total field" took the lead, and the terms "physical space" and "ether" were in turn relegated to second place. Actually, at first Einstein used the expression "ether" less and less frequently, until it nearly completely vanished. Its existence was sustained only thanks to the re-publishing of Einstein's two works on the relativistic ether. In the final period of Einstein's academic activity in the years 1950-1955, total field gained a privileged ontological status in his works, becoming a primary concept, as compared to space and time. Space and time simply became properties of the continuum, *i.e.*, of the total field.

> Space does not enter here as something existentially independent but as a continuous field of four dimensions.[455] **f22**

## 5.6    Should the expressions "new ether" and "relativistic ether" be used today?

As mentioned in Chapter 4, there is currently something of a revival of ether theories, and consequently also a revival of interest in Einstein's conception of ether. The use of the expressions "stationary ether," "new ether" and "relativistic ether" has already become common at the International Conferences on Physical Interpretations of Relativity Theory organised by M. C. Duffy every second year at Imperial College in London,[456] as well as during the International Conferences on Frontiers of Fundamental Physics, Olympia, Greece, 27-30 September 1993,[457] and "Relativistic Physics and Some of its Applications," Athens, Greece, 25-28 June 1997, organised by Michele Barone and Franco Selleri.[458] Once again we find the word "ether" in scientific articles and books; for example, in the book *The Philosophy of Vacuum* we read:

> Today the vacuum is recognised as a rich physical medium [...].A general theory of the vacuum is thus a theory of everything, a

---

[455] A. Einstein, "Prefazione di A. Einstein," in: *Cinquant'anni di Relatività 1905-1950* (Firenze: Giunti Barbèra, 1955), p. XVI.

[456] M. C. Duffy, (ed.) *Physical Interpretations of Relativity Theory*, British Society for Philosophy of Science, London, Proceedings of the International conferences of 1988, 1990, 1992, 1994, 1996 and 1998.

[457] M. Barone and F. Selleri eds., *Frontiers of Fundamental Physics* (New York-London: Plenum Press, 1994).

[458] F. Selleri, ed., *Open Questions in Relativistic Physics* (Montreal: Apeiron, 1998).

universal theory. It would be appropriate to call the vacuum "ether" once again [...][459] **f23**

Let us first consider the problem of terminology from the philological point of view. It is strange, yet true, that in physics we can find several terms that no longer correspond to their original etymological meaning, even though they are still used: *e.g.* "atom," "vacuum." There are still other words that have not lost their most fundamental meaning but have been rejected, *e.g.* "ether." The word "atom" no longer means an indivisible corpuscle, and a return to its original meaning seems quite impossible, because atoms, in the modern sense of the word, are composed of elementary particles, and nucleons, being the components of atoms, seem to consist of quarks, which are regarded as even more fundamental elementary particles. We are in a similar situation when we use the term "vacuum," which has also lost its original meaning. The "vacuum" is no longer a real vacuum and, therefore, Einstein was of the opinion that it should be called a "plenum,"[460] because it constitutes a *sui generis* fundamental material medium possessing a real structure and other physical qualities. Since the vacuum has real physical properties and (as Einstein recognised), as a fundamental field it is a certain distribution of energy, there is one good reason to call it "new ether." The much-maligned word "ether," which has always meant a certain specific kind of matter, is best suited, from the philological point of view, to express the four-dimensional space-time continuum of relativity theory. Einstein used it without any hesitation when he became aware that the space-time of his relativity theory had real physical properties and did not constitute a real vacuum. The return to the word "ether" is fully possible and justified here. It is justified both from the philological point of view and for other reasons: historical, didactic, and physical.

Modern science has its roots in ancient Greek philosophy. This philosophy, as we know, used the word "ether" to designate the particular kind of matter that filled the universe. This term was used throughout the history of philosophy and science, and it was also current at the beginning of this century. A resumption of its use at the dawn of this new century is now a fact. Since, according to General Theory of Relativity and other modern branches of physics, the space and time of

---

[459] S. Saunders and H. R. Brown, eds., *The Philosophy of Vacuum* (Oxford: Clarendon Press, 1991), p.251.

[460] A. Einstein, "Relativity and the Problem...," pp. 375-376.

the universe do not constitute a *vacuum*, but a structured material *plenum* characterised by different physical quantities, the historical and traditional word "ether" is the most appropriate to express these features of the universe.

It might be claimed that a return to Einstein's concept of a "new ether" would create a certain confusion in terminology, because in the minds of physicists (and not just physicists) the expression "ether" is closely linked with the notions and concepts of nineteenth century physics. Hence, it would not be advisable to use it from the didactic point of view, because for many it would mean a reversion, in one way or another, to the concepts of the past. But this is certainly not true, because a good teacher would be able to differentiate these concepts and teach all the different models of the ether using different names and adjectives, *e.g.*, Einstein's ether, Lorentz's ether, Weyl's ether, Eddington's ether, Dirac's ether and "new ether," "relativistic ether," stationary ether, nineteenth century physics ether, and so on—each in its proper context. Consequently, everything would be clear for students, and there is no reason why the expression "ether" should not be used from the didactic point of view. On the contrary; for precisely this reason it is preferable to use a traditional word, because it fully expresses the particular kind of materiality of the space-time continuum. When we use only the expression "space-time continuum," its materiality is not indicated in any way. We therefore need a special word to express it. The traditional word "ether" is ideally suited to this purpose.

The expressions "new ether" and "relativistic ether" are particularly useful from the physical point of view because they indicate immediately that in the Theory of Relativity, space-time is of a material nature. It is well to recall the reasons why Einstein attributed material properties to the space-time continuum:

1. The space-time continuum participates, in a real and active way, in physical becoming. *E.g.* the gravitational potentials described mathematically by the $g_{\mu\nu}$ components of the metrical tensor $g$ determine the inertio-gravitational behaviour of test particles.
2. Space-time is a field, and there is no qualitative difference between field and matter. Field is characterised by a certain distribution of energy, and therefore, materiality.

Note that, after 1916 and until his death, Einstein was never against the expression "new ether." He used this term frequently in his scientific correspondence with scientists, and in his scientific interpretative papers until 1938. After 1938 he did not write any new interpretative papers on the ether, but he authorised re-releases of his Leiden lecture on the new ether and other papers about the relativistic ether. In one of them, as we know, he introduced some amendments. He instructed the publisher to omit the part that was no longer in accordance with his opinions, but he did not remove the idea of the new ether. In the revised version of the paper, the expression "new ether" acquired an even more general meaning.

It is fitting to close this essay with two quotes from Einstein that show in which meaning the word "ether" can be used in his relativity theory, and why we cannot do without the ether in theoretical physics.

> We may still use the word *ether* but only to express the physical properties of space. The word *ether* has changed its meaning many times in the development of science. At the moment, it no longer stands for a medium built up of particles. Its story, by no means finished, is continued by the relativity theory.[461] **f24**

> [...] we will not be able to do without the ether in theoretical physics, *i.e.*, a continuum which is equipped with physical properties; for the general theory, whose basic points of view physicists surely will always maintain, excludes direct distant action. But every contiguous action theory presumes continuous fields, and therefore also the existence of an "ether."[462] **f25**

---

[461] A. Einstein, L. Infeld, *Die Physik als Abenteuer der Erkenntnis* (Leiden: A. W. Sijthoff, 1949), pp.99-100. See Introduction, footnote 1.

[462] A. Einstein, "Über den Äther," p. 93. The English translation in: S. Saunders and H.R. Brown, eds., *The Philosophy of Vacuum* (Oxford: Clarendon Press, 1991), p. 20.

# Appendix

# ORIGINAL QUOTATIONS

## a: Introduction

**a1**    "Wir mögen noch weiterhin das Wort Äther gebrauchen, jedoch nur, um dadurch die physikalischen Eigenschaften des Raums auszudrücken. Dieses Wort Äther hat in der Entwicklung der Wissenschaft viele Male seine Bedeutung geändert. Gegenwärtig bezeichnet es nicht mehr ein Medium, das irgendwie aus materiellen Partikeln aufgebaut ist. Seine Geschichte ist aber noch keineswegs beendet und wird durch die Relativitätstheorie fortgesetzt werden." A. Einstein and L. Infeld, *Die Physik als Abenteuer der Erkenntnis* (Leiden: A.W. Sijthoffs Witgeversmaatschappij N.V., 1949), pp. 99-100.

**a2**    "Deshalb war ich in 1905 der Ansicht, dass man von dem Äther in der Physik überhaupt nicht mehr sprechen dürfe. Dieses Urteil war aber zu radical, wie wir bei den folgenden Überlegungen über die allgemeine Relativitätstheorie sehen werden. Es bleibt vielmehr nach wie vor erlaubt, ein raumerfüllendes Medium anzunehmen als dessen Zustände man die elektromagnetischen Felder (und wohl dann auch die Materie) ansehen kann." A. Einstein, "Grundgedanken und Methoden der Relativitätstheorie in ihrer Entwicklung dargestellt," (*Morgan Manuscript*), *EA* 2070.

**a3**    "Es wäre richtiger gewesen, wenn ich in meinen früheren Publikationen mich darauf beschränkt hätte, die Nichtrealität der Äther*geschwindigkeit* zu betonen, statt die Nicht-Existenz des Äthers überhaupt zu vertreten. Denn ich sehe ein, dass man mit dem Worte Äther nichts anderes sagt, als dass der Raum als Träger physikalischer Qualitäten aufgefasst werden muss." A. Einstein, Letter to H.A. Lorentz, 15/11/1919, *EA*, 16, p.494.

# b: Chapter 1

**b1**  "Vor allem aber muss sich zeigen lassen, dass es für den elektrischen Strom zur Bildung des magnetischen Feldes einen passiven Widerstand gibt, der proportional ist der Länge der Strombahn und unabhängig vom Querschnitt und Material des Leiters." A. Einstein, "Über die Untersuchung des Ätherzustandes im magnetischen Felde," *PhB*, 27 (1971), pp. 390-391. The quoted sentence is on p. 391.

**b3**  "Zur Erforschung der Relativbewegung der Materie gegen den Lichtaether ist mir wieder eine erheblich einfachere Methode in den Sinn gekommen, welche auf gewöhnlichen Interferenzversuchen beruht. Wenn mir nur einmal das unerbittliche Schicksal die zur Ausführung nötige Zeit und Ruhe gibt! Wenn wir uns wieder einmal sehen, werde ich Dir darüber berichten." A. Einstein, Letter to M. Grossmann, 6/9/1901, *EA*, 11-485.

**b4**  "Es wird mir immer mehr zur Überzeugung, dass die Elektrodynamik bewegter Körper, wie sie sich gegenwärtig darstellt, nicht der Wirklichkeit entspricht, sondern einfacher wird darstellen lassen. Die Einführung des Namens "Äther" in die elektrischen Theorien hat zur Vorstellung eines Mediums geführt, von dessen Bewegung man sprechen könne, ohne dass man wie ich glaube, mit dieser Aussage einen physikalischen Sinn verbinden kann." A. Einstein, Letter to M. Marić, July 1899?, *EA*, FK-53.

**b5**  "Wir machen nun die Hypothese, dass der Aether stets vollständig in Ruhe bleibt. Auf diese Grundlage hat H.A. Lorentz eine sehr vollständige und elegante Theorie entwickelt, welche der hier gegebenen Darstellung im Wesentlichen zu Grunde liegt. Die Vorstellung des absolut ruhenden Aethers ist an sich schon die einfachste und natürlichste, wenn man nämlich unter dem Aether nicht eine Substanz, sondern lediglich den mit gewissen physikalischen Eigenschaften ausgestatteten Raum versteht." P. Drude, *Lehrbuch der Optik* (Leipzig: Verlag S. Hirzel, 1900), p. 419-420.

**b7**  "Gerade so gut, wie man einem besonderem Medium, welches den Raum überall erfüllt, die Vermittlerrolle von Kraftwirkungen zuweist, könnte man auch dasselbe entbehren und dem Raum selbst diejenigen physikalischen Eigenschaften beilegen, welche dem Aether jetzt zugeschrieben werden. Man hat sich bisher vor dieser Anschauung gescheut, weil man mit dem Worte "Raum" eine abstrakte Vorstellung ohne physikalische Eigenschaften verbindet. Da die Einführung des neuen Begriffes "Aether" durchaus ohne Belang ist, wofern man nur das Princip der Nahekräfte festhält, so soll in dieser Darstellung von der bisher üblichen Bezeichnung, d. h. der Einführung des Wortes "Aether," Gebrauch gemacht werden." P. Drude, *Physik des Aethers auf elektro-magnetischer Grundlage* (Stuttgart: Verlag von F. Henke, 1894), p. 9.

**b8**  "Das Wort "Aether" schliesst dabei keine neue Hypothese ein, sondern es ist nur der Inbegriff des von Materie freien Raumes, welcher gewisse physikalische Eigenschaften besitzt." P. Drude, *Die Theorie in der Physik* (Leipzig: Verlag S. Hirzel, 1895), p. 9.

**b9** "In einem Brief vom 8. April 1952 schrieb Einstein an Carl Seelig: "Auf Ernst Machs *Die Mechanik in ihrer Entwicklung* wurde ich von meinem Freunde Besso als Student, etwa im Jahre 1897, aufmerksam gemacht. Das Buch hat mit seiner kritischen Einstellung zu den Grundbegriffen und Grundgesetzen einen tiefen und nachhaltigen Eindruck auf mich ausgeübt." A. Einstein, Letter to C. Seelig, 8/4/1952, quoted by: G. Holton, *Thematische Analyse der Wissenschaft* (Frankfurt a. M.: Suhrkamp Verlag, 1981), p. 208.

**b9a** "[...]so könnte man doch hoffen, über dieses hypothetische Medium in Zukunft mehr zu erfahren, und sie wäre naturwissenschaflich noch immer wertvoller als der verzweifelte Gedanke an den absoluten Raum." E. Mach, *Die Mechanik in ihrer Entwiklung historisch-kritisch dargestellt*, 8 Aufl. (Leipzig: Brockhaus, 1920), p. 225.

**b10** "Dies spräche für den Äther als Medium der Schwere." E. Mach, *Die Mechanik in ihrer Entwicklung historisch-kritisch dargestellt*, 8 Aufl. (Leipzig: Brockhaus, 1920), p.186

**b11** "Ich sehe Machs wahre Grösse in der unbestechlichen Skepsis und Unabhängigkeit; in meinen jungen Jahren hat mich aber auch Machs erkenntnistheoretische Einstellung sehr beeindruckt, die mir heute als im Wesentlichen unhaltbar erscheint. Er hat nämlich die dem Wesen nach konstruktive und spekulative Natur alles Denkens und im Besonderen des wissenschaftlichen Denkens nicht richtig ins Licht gestellt und infolge davon die Theorie gerade an solchen Stellen verurteilt, an welchen der konstruktiv-spekulative Charakter unverhüllbar zutage tritt." A. Einstein, "Autobiographisches (1946)," [in:] *Albert Einstein als Philosoph und Naturforscher*, P.A. Schilpp ed. (Braunschweig/Wiesbaden: Friedr. Vieweg und Sohn, 1979), pp.1-35. The quoted sentences are on p. 8.

**b12** "Im Interesse einer möglichst hypothesenfreien Naturauffassung ist zu fragen, ob die Annahme jenes Mittels, des Äthers, unvermeidlich ist. Mir scheint das nicht der Fall zu sein.

"Ich kann nicht unternehmen, die vorstehend gemachten Andeutungen zu einer vollständigen Lichttheorie zu entwickeln; es kam mir nur darauf an, auf die Möglichkeit einer rein energetischen Behandlung derselben hinzuweisen. Der Hauptpunkt dabei ist, dass, nachdem die Energie als ein reales Wesen, ja das einzige reale Wesen der sogenannten Aussenwelt erkannt ist, wir kein Bedürfnis mehr haben, nach einem Träger derselben zu suchen, wenn wir sie irgendwo antreffen. Dies ermöglicht uns, die strahlende Energie als selbständig im Raume bestehend anzusehen." W. Ostwald, *Lehrbuch der allgemeinen Chemie* (Leipzig: W. Engelmann, 1893), Bd. 2, 1Tl., pp. 1014 and 1016.

**b13** "*La science et l'hypothese* de Poincaré, un livre qui nous a profondément impressionés et tenus en haleine pendant de longues semaines." Maurice Solovine, *Introduction*, in: A. Einstein, *Briefe an Maurice Solovine (1906-1955)* (Berlin: Veb deutscher Verlag der Wissenschaft, 1960), p. X.

**b14** "Mileva, intelligente et réservée, nous écoutait attentivement, mais n'intervenait jamais dans nos discussions." Maurice Solovine, "Introduction," in: A. Einstein,

*Briefe an Maurice Solovine (1906-1955)* (Berlin: Veb deutscher Verlag der Wissenschaft, 1960), p. XIV.

## c: Chapter 2

c1    "Diese als Bedingung des dynamischen Gleichgewichtes gefundene Beziehung entbehrt nicht nur der Übereinstimmung mit der Erfahrung, sondern sie besagt auch, dass in unserem Bilde von einer bestimmten Energieverteilung zwischen Äther und Materie nicht die Rede sein kann." A. Einstein, "Über einen die Erzeugung und Verwandlung des Lichtes betreffenden heuristischen Gesichtspunkt," *AdP*, 17 (1905), pp. 132-148.

c3    "Beispiele ähnlicher Art, sowie die misslungenen Versuche, eine Bewegung der Erde relativ zum "Lichtmedium" zu konstatieren, führen zu der Vermutung, dass dem Begriffe der absoluten Ruhe nicht nur in der Mechanik, sondern auch in der Elektrodynamik keine Eigenschaften der Erscheinungen entsprechen." A. Einstein, "Zur Elektrodynamik bewegter Körper," *AdP*, 17 (1905), pp. 891-921.

c4    "D'un autre côté, on a fait bien des recherches sur l'influence du mouvement de la terre. Les résultats ont toujours été négatifs." H. Poincaré, *La science et l'hypothèse* (Paris: Flammarion, 1968), p. 182.

c5    "Dass die Elektrodynamik Maxwells—wie dieselbe gegenwärtig aufgefasst zu werden pflegt—in ihrer Anwendung auf bewegte Körper zu Asymmetrien führt, welche den Phänomenen nicht anzuhaften scheinen, ist bekannt." A. Einstein, "Zur Elektrodynamik bewegter Körper," *AdP*, 17 (1905), pp. 891-921. The quoted sentence is on p. 891.

c6    "Einführung eines "Lichtäthers" wird sich insofern als überflüssig erweisen, als nach der zu entwickelnden Auffassung weder ein mit besonderen Eigenschaften ausgestatteter "absolut ruhender Raum" eingeführt, noch einem Punkte des leeren Raumes, in welchem elektromagnetische Prozesse stattfinden, ein Geschwindigkeitsvektor zugeordnet wird." A. Einstein, "Zur Elektrodynamik bewegter Körper," *AdP*, 17 (1905), pp. 891-921. The quoted sentence is on page 892.

c7    "Wenn die Theorie den Tatsachen entspricht, so überträgt die Strahlung Trägheit zwischen den emittierenden und absorbierenden Körpern." A. Einstein, "Ist die Trägheit eines Körpers von seinem Energieeinhalt abhängig? *AdP*, 18 (1905), pp. 639-641. The quoted sentence is on p. 641.

c8    "Nur die Vorstellung eines Lichtäthers als des Trägers der elektrischen und magnetischen Kräfte passt nicht in die hier dargelegte Theorie hinein; elektromagnetische Felder erscheinen nämlich hier nicht als Zustände irgendeiner Materie, sondern als selbstständig existierende Dinge, die der ponderabeln Materie gleichartig sind und mit ihr das Merkmal der Trägheit gemeinsam haben." A. Einstein, "Relativitätsprinzip und die aus demselben gezogenen Folgerungen," *JR*, 4 (1907), pp. 411-462. The quoted sentence is on p. 413.

**c9** "[...] allgemeiner überhaupt relativ zu jedem beschleunigungsfrei bewegten System nach genau den gleichen Gesetzen verlaufen. Diese Voraussetzung wollen wir im folgenden kurz "Relativitätsprinzip" nennen. Bevor wir die Frage berühren, ob es möglich sei, an dem Relativitätsprinzip festzuhalten, wollen wir kurz überlegen was bei Festhaltung dieses Prinzips aus der Ätherhypothese wird.

"Unter Zugrundelegung der Ätherhypothese führte das Experiment dazu, den Äther als unbeweglich anzunehmen. Das Relativitätsprinzip besagt dann, dass alle Naturgesetzte in bezug auf ein relativ zum Äther gleichförmig bewegtes Koordinatensystem $K'$ gleich seien den entsprechenden Gesetzen in bezug auf ein relativ zum Äther ruhendes Koordinatensystem $K$. Ist dem aber so, dann haben wir ebensoviel Grund, uns den Äther als relativ zu $K'$ ruhend vorzustellen wie als relativ zu $K$ ruhend. Es ist dann überhaupt ganz unnatürlich, eines der beiden Koordinatensysteme $K$, $K'$ dadurch auszuzeichnen, dass man einen relativ zu ihm ruhenden Äther einführt. Daraus folgt, dass man zu einer befriedigenden Theorie nur dann gelangen kann, wenn man auf die Ätherhypothese verzichtet. Die das Licht konstituirenden elektromagnetischen Felder erscheinen dann nicht mehr als Zustände eines hypothetischen Mediums, sondern als selbständige Gebilde, welche von den Lichtquellen ausgesandt werden, gerade wie nach der Newtonschen Emissiontheorie des Lichtes. Ebenso wie gemäss letzerer Theorie erscheint ein nicht von Strahlung durchsetzer, von ponderabler Materie freier Raum wirklich als leer." A. Einstein, "Entwicklung unserer Anschauungen über das Wesen und die Konstitution der Strahlung," *PhZ*, 10 (1909), pp 817-825. The quoted sentence is on p. 819.

**c10** "Le premier pas à faire si l'on veut tenter une telle conciliation, *c'est de renoncer à l'éther.*" A. Einstein, "Principe de la relativité et ses consequences dans la physique moderne," *ASPN*, 29 (1910), pp. 5-28 and 125-244. The quoted sentence is on p. 18.

**c11** "Dass wir ferner auf die Einführung eines Lichtäthers in die Theorie zu verzichten haben, ist leicht einzusehen. Denn wenn jeder Vakuumlichtstrahl sich in bezug auf $K$ mit der Geschwindigkeit $c$ fortpflanzen soll, so müssen wir jenen Lichtäther als in bezug auf $K$ überall ruhend denken. Wenn aber die Gesetze der Lichtfortpflanzung in bezug auf das (relativ zu $K$ bewegte) System $K'$ dieselben sind, wie in bezug auf $K$, so müssten wir mit demselben Rechte die Existenz eines in bezug auf $K'$ ruhenden Lichtäthers annehmen. Da es absurd ist, anzunehmen, der Lichtäther ruhe gleichzeitig in bezug auf beide Systeme, und da es kaum minder absurd wäre, in der Theorie eines der beiden (bzw. unendlich vielen) physikalisch gleichwertigen Systeme vor der anderen auszuzeichnen, so muss man auf die Einführung jenes Begriffes verzichten, der ohnehin nur nutzloses Beiwerk der Theorie war, seitdem man auf eine mechanische Deutung des Lichtes verzichtet hatte." A. Einstein, "Relativitätstheorie," [in:] *Die Physik*, unter Redaktion von E. Lecher (Leipzig: Teubner, 1915), pp. 702-713. The quoted sentence is on p. 708.

**c12**  "Die im Folgenden skizzierte Theorie ist mit der Äther-Hypothese nicht vereinbar." A. Einstein, "Relativitätstheorie," *VNGZ*, 56 (1911), pp. 1-14. The quoted sentence is on p. 2.

**c13**  "Mit den von Hrn. Lenard beobachteten Eigenschaften der lichtelektrischen Wirkung steht unsere Auffassung, soweit ich sehe, nicht im Widerspruch." A. Einstein, "über einen die Erzeugung und Verwandlung des Lichtes betreffenden heuristischen Gesichtspunkt," *AdP*, 17 (1905), pp. 132-148. The quoted sentence is on p. 147.

**c14**  "Lieber Herr Laub! Zuerst meine herzlichste Gratulation wegen der Assistentur und des damit verbundenen Einkommens. Ich hatte mein grosses Vergnügen an dieser Nachricht. Aber ich glaube, dass die Gelegenheit, mit Lenard zusammen zu arbeiten, noch weit mehr ist als Assistentur und Einkommen zusammen! Ertragen Sie Lenards Schrullen, soviel er nur haben mag. Er ist ein grosser Meister, ein origineller Kopf! Vielleicht ist er ganz gut umgänglich einem Mann gegenüber, den er achten gelernt hat." A. Einstein, Letter to J.J. Laub (1908) [in:] A. Kleinert, Ch. Schönbeck, "Lenard und Einstein. Ihr Briefwechsel und ihr Verhältnis vor der Nauheimer Diskussion,"*Gesnerus*, 35 (1978), pp. 318-333. The quoted sentence is on p. 320.

**c15**  "Und doch müssen Sie sich glücklich preisen, dass Sie bei Lenard sind, zumal Sie ja—wie es scheint—ihn mit grossem Geschick zu behandeln verstehen. Er ist nicht nur ein geschickter Meister in seiner Zunft, sondern wirklich ein Genie." A. Einstein, Letter to J.J. Laub, 16/3/1910; [in:] A. Kleinert, Ch. Schönbeck, "Lenard und Einstein. Ihr Briefwechsel und ihr Verhältnis vor der Nauheimer Diskussion," *Gesnerus* 35 (1978), pp. 318-333. The quoted sentence is on p. 320.

**c16**  "Lenard muss aber in vielen Dingen sehr "schief gewickelt" sein. Sein Vortrag von neulich über die abstruse Ätherei erscheint mir fast infantil." A. Einstein, Letter to J.J. Laub, date unknown, [in:] A. Kleinert, Ch. Schönbeck, "Lenard und Einstein. Ihr Briefwechsel und ihr Verhältnis vor der Nauheimer Diskussion," *Gesnerus* 35 (1978), pp. 318-333. The quoted sentence is on p. 322.

**c17**  "[...] eine ordentliche Professor für theoretische Physik zu errichten, 'wenn eine Persönlichkeit wie Einstein [...] dafür zur Verfügung stünde. Ph. Lenard, Letter to A. Sommerfeld, 4/9/1913, in: A Kleinert and Ch. Schönbeck, "Lenard und Einstein. Ihr Briefwechsel und ihr Verhältnis vor der Nauheimer Diskussion," *Gesnerus* 35 (1978), p. 322.

**c18**  "M. H.! Die Anschauungen über Raum und Zeit, die ich Ihnen entwickeln möchte, sind auf experimentell-physikalischem Boden erwachsen. Darin liegt ihre Stärke. Ihre Tendenz ist eine radikale. Von Stund an sollen Raum für sich und Zeit für sich völlig zu Schatten herabsinken und nur noch eine Art Union der beiden soll Selbständigkeit bewahren." H. Minkowski, "Raum und Zeit," *PhZ*, 10 (1909), p. 104-111.

**c19**  "Doch ich bin ganz sicher, dass wir in diesem Seminar alles durchsprachen, was zu der Zeit über Elektrodynamik und Optik bewegter Systeme bekannt war. Wir studierten die Arbeiten von *Hertz, Fitzgerald, Larmor, Lorentz, Poincaré* und anderen.

Darüber hinaus aber erhielten wir einen Einblick in *Minkowskis* eigene Gedanken, die erst zwei Jahre später veröffentlich wurden. *Minkowski* veröffentlichte seine Arbeit 'Die Grundlagen für die elektromagnetischen Vorgänge in bewegten Körpern' im Jahre 1907. Sie enthält eine systematische Darstellung seiner formalen Verschmelzung von Raum und Zeit zu einer vierdimensionalen 'Welt' mit einer pseudo-euklidischen Geometrie, für die er eine Vektor- und Tensor-Rechnung entwickelte." M. Born, "Physik und Relativität," [in:] M. Born, *Physik im Wandel meiner Zeit*, Fourth edn. (Braunschweig: F. Vieweg u. Sohn, 1966), pp. 186 and 192.

c20 "Das Prinzip von der Konstanz der Lichtgeschwindigkeit gilt nach dieser Theorie nicht in derjenigen Fassung, wie es der gewöhnlichen Relativitätstheorie zugrunde gelegt zu werden pflegt." A. Einstein, "Über den Einfluss der Schwerkraft auf die Ausbreitung des Lichtes," *AdP*, 35 (1911), pp. 898-908. The quoted sentence is on p. 906.

c21 "Man kann bei dieser Auffassung ebensowenig von der *absoluten Beschleunigung* des Bezugssystems sprechen, wie man nach der gewöhnlichen Relativitätstheorie von der *absoluten Geschwindigkeit* eines Systems reden kann." A. Einstein, "Über den Einfluss der Schwerkraft auf die Ausbreitung des Lichtes," *AdP*, 35 (1911), pp. 898-908. The quoted sentence is on p. 899.

c22 "[...] wenigstens hat meiner Meinung nach die Hypothese, dass das 'Beschleunigungsfeld' ein Spezialfall des Gravitationsfeldes sei, eine so grosse Wahrscheinlichkeit [...]" c55: A. Einstein, "Lichtgeschwindigkeit und Statik des Gravitationsfeldes," *AdP*, 38 (1912), pp. 355-369. The quoted sentence is on p. 355.

c23 "[...] in einem gleichförmig rotierenden Systeme, in welchem wegen der Lorentzkontraktion das Verhältnis des Kreisumfanges zum Durchmesser bei Anwendung unserer Definition für die Längen von $\pi$ verschieden sein müsste." A. Einstein, "Lichtgeschwindigkeit und Statik des Gravitationsfeldes," *AdP*, 38 (1912), pp. 355-369. The quoted sentence is on p. 356.

c26 **(Polish original)** "Praca jest dokładnie tym, co zapowiada w tytule -*projektem* nowej teorii grawitacji, która byłaby równocześnie uogólnieniem teorii względności z 1905. Sformułowanie tego projektu było niewątpliwie punktem przełomowym. Teraz stało się jasne, że wszystkie dotychczasowe wysiłki były tylko poszukiwaniem — często intuicyjnie i po omacku — nowych idei i próbami układania ich we fragmenty większej całości. Teraz wszystko nagle 'zaskoczyło' na swoje miejsce. Już było wiadomo, jaka będzie całość, chociaż nie zawsze jeszcze wiadomo, przy pomocy jakich narzędzi tę całość skonstruować. Dalszy ciąg wielkiej przygody będzie już tylko drogą — prawdą, pełną dramatycznych pomyłek i kroków wstecz — ale drogą, która wiadomo, dokąd ma prowadzić." M. Heller, "Jak Einstein stworzył ogólną teorię względności? (How did Einstein create the General Theory of Relativity?)" *PF*, 39 (1988), pp. 3-21. The quoted passage is on pp. 8-9.

**c27** "Für mich ist es absurd, dem 'Raum' physikalische Eigenschaften zuzuschreiben. Die Gesamtheit der Massen erzeugt ein $g_{\mu\nu}$-Feld (Gravitationsfeld), das seinerseits den Ablauf aller Vorgänge, auch die Ausbreitung der Lichtstrahlen und das Verhalten der Massstäbe und Uhren regiert. Das Geschehen wird zunächst auf vier *ganz willkürliche* raumzeitliche Variable bezogen." A. Einstein, Letter to E. Mach, undated, [v:] V.P. Vizgin and Ya.A. Smorodinskii, "From the equivalence principle to the equation of gravitation," *Sov. Phys. Usp.*, 22 (1979), pp. 489-515. The statement is on p. 499.

**c28** "Dziś jest to sprawa podręcznikowa (ale nadal stanowiąca poważną trudność dla początkujących adeptów teorii względności), Einstein i Grossmann nie wiedzieli o tożsamościach Bianchiego i o tym, że spośród dziesięciu składowych zaproponowanych przez nich równań pola tylko sześć może zawierać treść fizyczną, a cztery odzwierciedlają jedynie swobodę wyboru układu współrzędnych i mogą być w zasadzie dowolnie wybrane. Nic dziwnego, że Einstein i Grossmann 'wykazali,' iż składowe tensora metrycznego (czyli potencjały grawitacyjne) nie mogą 'być zdeterminowane' przez równania, bo istotnie nie mogą; cztery składowe można przecież wybrać dowolnie. Tkwi w tym pewien paradoks: 'Niedeterminiwanie' również wynika z ich niezmienniczości (niezależności od wyboru układu współrzędnych). Einstein poszukiwał równań niezmienniczych, a gdy je znalazł, to odrzucił właśnie dlatego (oczywiście nie wiedząc o tym), że posiadały własność, która jest następstwem niezmienniczości. Dalszy ciąg poszukiwań Einsteina—bo względny sukces artykułu z Grossmannem dał mu zadowolenie na krótko—będzie polegał w gruncie rzeczy na tym, by zrozumieć błąd popełniony w 1913 roku." M. Heller, "Jak Einstein stworzył ogólną teorię względności? (How did Einstein create the General Theory of Relativity?)" *PF*, 39 (1988), pp. 3-21. The quoted passage is on pp. 11-15.

**c29** "Im Weltraum schweben in grosser Entfernung von allen Himmelskörpern zwei Massen. Dieselben seien einander nahe genug, um Wirkungen aufeinander ausüben zu können. Ein Beobachter verfolge nun die Bewegung beider Körper, indem er stets in Richtung der Verbindungslinie beider Massen nach dem Fixsterngewölbe visiert. Er wird wahrnehmen, dass die Visierlinie am sichtbaren Fixsterngewölbe eine geschlossene Linie herausschneidet, welche ihren Ort in bezug auf das sichtbare Fixsterngewölbe nicht verändert. Wenn der Beobachter natürlichen Verstand besitzt, aber weder Geometrie noch Mechanik gelernt hat, so wird er so schliessen: 'Meine Massen führen eine Bewegung aus, welche wenigstens zum Teil vom Fixstern-System kausal bestimmt wird. Die Gesetze, nach denen sich Massen in meiner Umgebung bewegen, werden mitbestimmt durch die Fixsterne.' Ein Mann, der durch die Schule der Wissenschaft gegangen ist, wird über die Einfalt unseres Beobachters lächeln und ihm sagen: 'Die Bewegung Deiner Massen hat mit dem Fixstern-Himmel nichts zu schaffen; sie wird vielmehr ganz unabhängig von den übrigen Massen durch die Gesetze der Mechanik bestimmt. Es gibt einen Raum $R$, in dem diese Gesetze gelten. Diese Gesetze sind so, dass Deine Massen fortgesetzt in einer Ebene dieses Raumes bleiben. Das Fixstern-System aber kann in diesem Raum nicht rotieren, weil es sonst durch gewaltige Zentrifugalkräfte zerrissen würde. Es ruht notwendigerweise (wenigstens beinahe!), wenn es überhaupt dauernd soll existieren können; daher kommt es, dass die Ebene in der sich Deine Masse bewegen, immer durch dieselbe Fixsterne hindurchgeht.'—Unser furchtloser Beobachter wird aber sagen: 'Du magst ja unvergleichlich gelehrt sein. *Aber ebensowenig, als ich je dazu zu bringen war, an Gespenster zu glauben, glaube ich an das riesige Ding, von dem Du mir sprichst, und das Du Raum nennst. Ich kann weder so etwas sehen, noch mir etwas darunter denken*[Italics—L.K.]. Oder soll ich mir Deinen Raum R als sehr subtiles Körpernetz denken, auf das sich die übrigen Dinge beziehen? Dann kann ich mir ausser R noch ein zweites solches Netz R' denken, das relativ zu R beliebig bewegt ist (z. B. rotiert). Gelten Deine Gleichungen dann auch zugleich relativ zu R?' Der gelehrte Mann verneint dies mit Sicherheit. Hierauf der Einfältige: 'Woher wissen denn aber die Massen, bezüglich welches der "Räume" R, R' etc. sie sich Deinen Gesetzen gemäss bewegen sollen, woran erkennen sie den Raum bezw. die Räume, nach dem sie sich zu richten haben?' Nun ist unser gelehrter Mann in grösster Verlegenheit. Er betont zwar, dass es derartige privilegierte Räume geben müsse, aber er weiss keinen Grund dafür anzugeben, warum jene Räume vor anderen ausgezeichnet sein könnten. Hierauf der Einfältige: '*Dann, halte ich bis auf Weiteres Deine bevorzugten Räume für müssige Erfindung und bleibe meiner Auffassung, dass das Fixtsterngewölbe das mechanische Verhalten meiner Versuchsmassen mitbestimmt.*' [Italics—L.K.]" A. Einstein, "Zum Relativitätsproblem," *Scientia*, 15 (1914), pp. 344-345.

c30 **(Original in Gothic script)** "Von den Hauptergebnissen der Relativitätstheorie seien hier zwei erwähnt, die auch den Laien interessieren müssen. Das erste derselben liegt darin, dass die Hypothese von der Existenz eines raumerfüllenden, der Lichtfortpflanzung dienenden Mediums, des Lichtäthers, fallen gelassen werden muss. Das Licht erscheint nach dieser Theorie nicht mehr als Bewegungszustand eines unbekannten Trägers, sondern als physikalisches Gebilde, dem eine durchaus selbstständige Existenz zuzuschreiben ist. Zweitens ergibt die Theorie, dass die Trägheit eines Körpers keine absolut unveränderliche Konstante ist, sondern mit dem Energie-Inhalte wächst. Die wichtigen Erhaltungssätze von der Masse und von der Energie verschmelzen so zu einem einzigen Satze; die Energie eines Körpers ist zugleich bestimmend für die Masse desselben." A. Einstein, "Vom Relativitätsprinzip," *Die Vossische Zeitung*, Nr. 209, 26/4/1914.

c31 "So gelangte ich zu der Forderung einer allgemeineren Kovarianz der Feldgleichungen zurück, von der ich vor drei Jahren, als ich zusammen mit meinem Freunde Grossmann arbeitete, nur mit schwerem Herzen abgegangen war." A. Einstein, "Zur allgemeinen Relativitätstheorie" (Postcript), *SPAW* (1915), 2. Teil, pp. 799-801. The quoted sentence is on p. 799.

c32 "In einer jüngst in diesen Berichten erschienenen Arbeit, habe ich Feldgleichungen der Gravitation aufgestellt, welche bezüglich beliebiger Transformationen von der Determinante 1 kovariant sind. In einem Nachtrage habe ich gezeigt, dass jenen Feldgleichungen allgemein kovariante entsprechen, wenn der Skalar des Energietensor der "Materie" verschwindet, und ich habe dargetan, dass die Einführung dieser Hypothese, durch welche Zeit und Raum der

letzten Spur objektiver Realität beraubt werden, keine prinzipiellen Bedenken
entgegenstehen." A. Einstein, "Erklärung der Perihelbewegung des Merkur aus
der allgemeinen Relativitättheorie," *SPAW* (1915), 2nd part, pp. 831-839. The
quoted sentence is on p. 831.

**c33**  "Das Relativitätpostulat in seiner allgemeinsten Fassung, welches die
Raumzeitkoordinaten zu physikalisch bedeutungslosen Parametern macht, führt
mit zwingender Notwendigkeit zu einer bestimmten Theorie der Gravitation."
A. Einstein, "Feldgleichungen der Gravitation," *SPAW* (1915), 2. Teil, pp. 844-
847. The quoted sentence is on p. 847.

**c34**  "Dass diese Forderung der allgemeinen Kovarianz, welche dem Raum und der
Zeit den letzten Rest physikalischer Gegenständlichkeit nehmen, eine natürliche
Forderung ist, geht aus folgender Überlegung hervor. Alle unsere zeiträumlichen
Konstatierungen laufen stests auf die Bestimmung zeiträumlichen Koinzidenzen
hinaus." A. Einstein, "Die Grundlagen der allgemeinen Relativitätstheorie," *AdP*,
49 (1916), pp. 769-822. The quoted sentence is on p. 774.

**c35**  "[...] indem die das Gravitationsfeld darstellenden 10 Funktionen $g_{\mu\nu}$ zugleich die
metrischen Eigenschaften des vierdimensionalen Messraumes bestimmen."
A. Einstein, "Die Grundlagen der allgemeinen Relativitätstheorie," *AdP*, 49
(1916), pp. 769-822. The quoted sentence is on p. 777.

## d: Chapter 3

**d1**  "Ich habe mich in den letzten Monaten viel mit Ihrer Gravitationstheorie und
allgemeiner Relativitätstheorie beschäftigt, und habe auch, was mir sehr nützlich
war, darüber vorgetragen. Ich glaube jetzt die Theorie in ihren vollen Schönheit zu
verstehen, jede Schwierigkeit auf die ich stiess, habe ich bei näherer Betrachtung
überwinden können, auch ist es mir gelungen, Ihre Feldgleichungen

$$G_{im} = -k \left( T_{im} - \frac{1}{2} g_{im} \, T \right)$$

aus dem Variationsprinzip abzuleiten, wenigstens fehlt an dieser Ableitung, die für
mich lange Rechnungen erforderte, nur noch eine Kleinigkeit.

"Ich bin nun aber auf eine Überlegung gekommen, die ich Ihnen vorlegen
möchte, und die auf der Betrachtung eines fiktiven Experiments beruht. Wir
können uns denken, dass man den Lecher'schen Versuch macht mit zwei
vollkommen leitenden Drähten, die am Aequator um die Erde herum angespannt
sind, und deren jeder in sich selbst geschlossen ist. Um der Gefahr des
"Entgleisens der elektromagnetischen Wellen" (wegen der Erdkrümung) zu
entgehen, können wir statt den beiden Drähten auch einen einzigen Draht mit
derselben konzentrier umgebenden leitender Hülle anwenden. An einer
bestimmten Stelle A dieses in sich geschlossenen "Kabels" (Raum zwischen den
Leitern luftleer) möge sich eine Vorrichtung befinden, die es ermöglicht Wellen zu
erregen, und ein Detektor, mit dem wir die nach Durchlaufung des Kreises in A

zurückkehrenden Wellen beobachten. Das Kabel sowie den Punkt A seien fest mit der Erde verbunden.

"Nach allem, was wir wissen, können wir wohl mit Bestimmheit sagen, was wir mit genügend verfeinerten Mitteln beobachten würden. Wellen, die in demselben Augenblick in A erzeugt werden, und den Kreis in entgegengesetzten Richtungen durchlaufen, werden *nicht* in demselben Augenblick in A zurückkehren.

"Unter den verschiedenen Weisen, auf die wir dieses Ergebnis beschreiben können, gibt es nur zwei, die besonders einfach sind.

"**a**. Wir können ein Koordinatensystem I OX, OY (OZ falle mit der Erdachse zusammen) so wählen, dass in diesem System die Fortpflanzungsgeschwindigkeit der Wellen für die beiden Umlaufrichtungen die gleiche ist. Wir finden dann, dass die Erde sich in dem Koordinatensystem dreht.

"**b**. Wir führen ein fest mit der Erde verbundenes Koordinatensystem II ein. In diesem bestehen für die beiden Umlaufsrichtungen ungleiche Fortpflanzungsgeschwindigkeiten $c_1$ und $c_2$.

"Es braucht kaum gesagt zu werden dass sich eben die nötige Verschiedenheit der Fortpflanzungsgeschwindigkeiten aus Ihren allgemeinen Formeln ergibt, wenn man von I zu II übergeht, und insofern eine Gleichung von der Gestalt

$$c_1 - c_2 = a$$

"sowohl im System I, wie in II und ebenso noch in vielen anderen System (jedesmal mit einem andern a) gilt, kann man sagen, sie drückt das Ergebnis des Versuches in *kovarianter* Form aus. Das braucht uns aber nicht davon abzuhalten, die Gleichung $c_1 - c_2 = 0$ als *verschieden* von $c_1 - c_2 \neq 0$ zu betrachten. In diesem Sinne werden wir schliessen: die Erscheinungen in dem Kabel spielen sich in Bezug auf die Koordinatensysteme I und II nicht in derselben Weise ab.

"Wenn man nun versucht, sich dies irgendwie verständlich zu machen oder bildlich vorzustellen, so wird man sich kaum darauf beschränken können, *nur* von der Erde, dem Kabel und dem in diesem letzteren enthaltenen "Raum" oder "Vakuum" zu sprechen, man wird ja geneigt sein sich vorzustellen, dass es in dem Raum oder dem Vakuum an und für sich nichts gibt, dass sich den Systemen I und II gegenüber verschieden verhält.

"Die Vorstellung liegt gewiss nahe [von einer anderen spreche ich weiter unten] und es hätte früher wohl allen Physikern sehr natürlich geschienen, dass es in dem Kabel ein Medium (Aether) gibt, *indem* sich die Wellen fortpflanzen, derart dass die Fortpflanzungsgeschwindigkeit relativ zum Medium immer dieselbe ist, dass aber dieses Medium in Bezug auf das eine Achsensystem ruhig, in Bezug auf das andere sich bewegen kann. Stellen wir uns auf diesen Standpunkt so können wir sagen, der Versuch habe uns die relative Bewegung der Erde gegen den Aether gezeigt. Haben wir dann in dieser Weise die Möglichkeit anerkannt, eine relative *Rotation* zu konstatieren, so dürfen wir nicht von vornherein die Möglichkeit leugnen, auch Aenderungen einer solchen *Translation* zu erhalten, d.h. wir dürfen den Grundsatz der Relativitätstheorie nicht als Postulat hinstellen. Wir müssen vielmehr (und das war auch der wirkliche Entwicklungsgang) die Beantwortung der Frage in den Beobachtungen suchen. Nachdem diese uns gelehrt haben dass ein Einfluss der Translation nicht gefunden werden kann, dürfen wir, indem wir (und zwar ziemlich weitgehend) generalisieren, jenen Satz als Grund*hypothese* aussprechen, wobei wir aber noch immer die Möglichkeit zulassen, (für wie wenig wahrscheinlich wir es auch halten mögen), dass künftige Beobachtungen uns zwingen werden, die Hypothese aufzugeben.

"Man kann diese Betrachtungen noch in anderer Weise einkleiden. Wir können nämlich in dem geschlossenen Kabel stehende Wellen erzeugen, und in jedem Augenblick die Lage der Knoten beobachten. Es wird sich dann ergeben, dass diese relativ zur Erde in Kreise herumlaufen. Man könnte sich nun allerdings darauf beschränken, die relative Bewegung der Knoten gegen die Erde (oder umgekehrt) zu konstatieren. Wenn man aber erwägt, dass dieselbe Rotation bei stehenden Wellen verschiedener Länge und verschiedener Intensität auftritt, so liegt es auf der Hand (sagen wir als bildliche Zusammenfassung des allen diesen Erscheinungen Gemeinsamen) an einen Aether zu denken, in welchem die stehenden Wellen ihren Sitz haben. Auch Mach, an dessen Auffassung Sie sich angeschlossen haben, hat bei der Besprechung ähnlicher Versuche das Bedürfnis empfunden, etwas ausserhalb der Erde liegendes, das für die Erscheinugen bestimmend wäre, anzunehmen. In seinem Gedankengange würde man in einem Einfluss der "entfernten Körper des Weltalls," sagen wir der Fixsterne, einen bestimmenden Moment suchen. Man würde also sagen, es sind die Fixsterne, welche das im Kreis herumlaufen (oder das ruhen) die Knoten in dem ring-förmigen Kabel bestimmen. Obgleich mir nun diese Auffassung viel weniger naheliegend scheint als die Hypothese eines Aethers, so könnte ich sie doch gelten lassen, wenn sie, im Vergleich mit dieser Hypothese irgend einen Vorteil böte. Aber einen solchen vermag ich nicht zu sehen. Wenn wir nämlich annehmen müssen, die Rotation der Erde in Bezug auf die Fixsterne, habe einen beobachtbaren Einfluss auf elektromagnetische Erscheinungen, so dürfen wir nicht von vornherein die Möglichkeit eines ähnlichen Einflusses einer *Translation* der Erde oder des Sonnensystems relativ zu den Fixsternen leugnen. Wir sind dann genau eben so weit wie mit der Aetherhypothese und wir haben experimentell zu untersuchen, ob vielleicht irgend eine Wirkung einer Translation besteht. Von einem Relativitäts*postulat* dürfte auch jetzt die Rede nicht sein.

"Übrigens sind die beiden Auffassungen, Einfluss der Fixsterne und Aetherhypothese im Grunde, wie mir scheint, nicht einmal weit voneinander verschieden. Gesetzt, ich nehme an, die Bewegung oder Ruhe der Knoten in unserem ringförmigen Kabel werde durch den Einfluss der Fixsterne bestimmt. Dann kann ich, um die Natur dieses Einflusses einigermassen festzulegen, in dem Kabel ein System starr mit einander verbundener Punkte gleichsam als Verbindungsglied zwischen Fixsternen und elektromagnetischen Wellen annehmen. Ich werde sagen, der besagte Einfluss äussere sich darin, dass die Knoten in Bezug auf dieses Punktsystem, das seinerseits mit den Fixsternen verbunden ist, feste Lagen haben. Von diesem Punktsystem zu einem Aether ist der Schritt nicht weit.

"Selbstverständlich geben auch andere Versuche, z. B. die von Ihnen und Mach besprochene zu ganz ähnlicher Betrachtung Anlass, und werden die vorstehenden Überlegungen Ihnen keineswegs neu sein. Der Hauptpunkt in denselben ist eigentlich, dass Abweichungen von der Relativitätstheorie auch nach der "Fixsternehypothese" sehr gut denkbar wären. Dass übringens sowohl die Relativitätstheorie wie auch Ihre Gravitationstheorie auch bei der von mir vertretenen Auffassung in vollem Umfange bestehen bleiben können brauche ich nicht zu sagen. Nur werden sie sich uns weniger als die einzig mögliche aufdrängen." H.A. Lorentz, Letter to A. Einstein, 6/6/1916, *EA*, 16-451.

Footnote by Lorentz: **Raum zwischen den Leitern luftleer.

**d2** "Nun zu Ihrer Interferenzbetrachtung! Es hat mich amüsiert, dass Sie genau auf dasselbe Beispiel verfallen sind, das auch ich mir in den letzten Jahren habe oft durch den Kopf gehen lassen. Ich gebe Ihnen zu, dass die allgemeine Relativitätstheorie der Aetherhypothese näher liegt als die spezielle Relativitätstheorie. Aber diese neue Aethertheorie würde das Relativitätsprinzip nicht mehr verletzen. Denn der Zustand dieses $g_{\mu\nu}$ = Aether wäre nicht der eines starren Körpers von selbständigem Bewegungszustande. Sondern ein Bewegungszustand wäre eine Funktion des Ortes, bestimmt durch die materiellen Vorgänge. Beispiel:

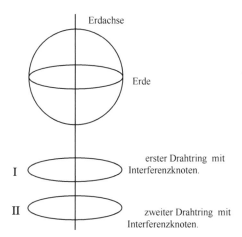

"Wäre die Erde nicht da oder würde sie sich nicht drehen, so würden die Interferenzknoten der Ringe I und II relativ zu den "Fixsternen," also auch relativ zu einander in Ruhe bleiben. Dreht sich die Erde aber, so drehen sich beide Knotensysteme aber in winzigem Prozentsatz mit, und zwar die von I wegen der geringeren Distanz mehr als die von II. Die Knotensysteme I und II rotieren also mit winziger Geschwindigkeit gegen einander nach Massgabe der Erddrehung und der Entfernungen. Die Foucaultsche Pendelebene dreht sich auch ein wenig mit der Erde, etwa 0,01" per Jahr. Schade, dass es nicht mehr ausmacht. Ich muss aber gestehen, dass mir das $g_{\mu\nu}$ System lieber ist als ein unvollkommener Vergleich mit einem stofflichen Etwas. Denn die Bevorzugung der gleichförmigen Bewegung findet in diesen modifizierten Aetherhypothesen keinen Ausdruck, wohl aber in dem abstrakten System. Geht man nämlich von einem Weltstück von konstanten $g_{\mu\nu}$ aus, so ändert eine lineare Substitution der $x_\nu$ nichts an der Kostanz der $g_{\mu\nu}$, wohl aber eine nicht lineare Substitution der $x_\nu$. Hieraus folgt, dass gleichförmige Relativbewegung kein Gravitationsfeld "erzeugt" d.h. unmerklich ist im Gegensatz zur ungleichförmigen Bewegung. Jener fundamentale Unterschied von gleichförmig und ungleichförmig kommt aber in der Aethervorstellung nicht unmittelbar zum Ausdruck; Man möchte vielmehr stets eine gleichförmige Bewegung nachweisen können." A. Einstein, Letter to H.A. Lorentz, 17/6/1916, *EA*, 16-453.

**d3** "Die Koeffizienten der metrischen Fundamentalform sind somit nicht bloss die Potenziale der Gravitations- und Zentrifugalwirkungen, sondern *bestimmen allgemein, welche Weltpunkte untereinander im Wirkungszusammenhang stehen.* Vielleich ist deshalb der Name "Gravitationsfeld" für dasjenige Reale, was durch diese Form dargestellt wird, zu einseitig und würde besser durch "*Äther*" ersetzt; während dann das elektromagnetische Feld schlechtwegs als Feld zu bezeichnen wäre. In der Tat spielt dieser "Äther" die gleiche Rolle wie der Äther der alten Lichttheorie und der "absolute Raum" der Newtonschen Mechanik; nur darf man nicht vergessen, dass er freilich ganz etwas anderes ist als ein substantieller Träger." H. Weyl, *Raum, Zeit, Materie* (Berlin: Verlag von Julius Springer, 1918), p. 182.

**d4** "Merkwürdig ist es dabei, dass gerade das verallgemeinerte Relativitätsprinzip, das mit besonderer Ausschliessungskraft dem Äther gegenüberzustehen scheint, zu "Raumkoordinaten" kommt, die diesem Prinzip wesentlich eigentümlich sind, die aber—die Variabilität ihrer Eigenschaften nach—sehr wohl als Bestimmungsstücke von Raumzuständen erscheinen können, wonach man geradezu den Eindruck empfängt, als ob hier eben der ausgeschlossene Äther unter dem veränderten Namen 'Raum' von selber sich wieder gemeldet hätte." Ph. Lenard, *Über Relativitätsprinzip, Äther, Gravitation* (Leipzig: Verlag von S. Hirzel, 1918), p. 28.

**d5** "Krit.: Wie steht es denn jetzt mit dem kranken Mann der theoretischen Physik, dem Äther, den manche von euch als entgültig tot erklärt haben?

"Rel.: Ein wechselvolles Schicksal hat er hinter sich, und man kann durchaus nicht sagen, dass er nun tot sei. Vor *Lorentz* existierte er als alles durchdringende Flüssigkeit, als gasähnliche Flüssigkeit und sonst noch in den verschiedensten Daseinformen, verschieden von Autor zu Autor. Mit *Lorentz* wurde er starr und verkörperte das "ruhende" Koordinatensystem bezw. einen bevorzugten Bewegungszustand in der Welt. Gemäss der speziellen Relativitätstheorie gab es keinen bevorzugten Bewegungszustand mehr; dies bedeutete Leugnung des Äthers im Sinne der früheren Theorien. Denn gab es einen Äther, so musste er in jedem Raum-Zeitpunkt einen bestimmten Bewegungszustand haben, der in der Optik eine Rolle spielen musste. Einen solchen bevorzugten Bewegungszustand aber gibt es nicht, wie die spezielle Relativitätstheorie lehrte, und darum gibt es auch keinen Äther in alten Sinne. Auch die allgemeine Relativitätstheorie kennt keinen bevorzugten Bewegungszustand in einem Punkte, den man etwa als Geschwindigkeit eines Äthers interpretieren könnte. Während aber nach der speziellen Relativitätstheorie ein Raumteil ohne Materie und ohne elektromagnetisches Feld als schlechthin leer, d. h. durch keinerlei physikalische Grössen charakterisiert erscheint, hat nach der allgemeinen Relativitätstheorie auch der in diesem Sinne leere Raum physikalische Qualitäten, welche durch die Komponenten des Gravitationpotentials mathematisch charakterisiert sind, welcher das metrische Verhalten dieses Raumteils sowie dessen Gravitationsfeld bestimmen. Man kann diesen Sachverhalt sehr wohl so auffassen, dass man von einem Äther spricht, dessen Zustand von Punkt zu Punkt stetig variiert. Nur muss man sich davor hüten, diesem 'Äther' stoffähnliche Eigenschaften (z. B. an jeder Stelle eine bestimmte Geschwindigkeit) zuzuschrieben." A. Einstein, "Dialog über Einwände gegen die Relativitättheorie," *Die Naturwissenschaften*, 6 (1918), pp. 701-702.

**d6** "(13) *Spezielle Relativitätstheorie und Aether.* Es ist klar, dass in der Relativitätstheorie die Vorstellung von einem ruhenden Aether keinen Platz hat. Sind nämlich die Systeme K und K' für die Formulierung der Naturgesetze vollkommen gleichwertig, so ist es inkonsequent, der Theorie einem Begriff zugrunde zu legen, der eines dieser Systeme vor den übrigen Systemen auszeichnet. Setzt man nämlich einen relativ zu K ruhenden Aether voraus, so ist der Aether relativ zu K' bewegt, was nicht zur physikalischen Gleichwertigkeit beider Systeme passt.

"Deshalb war ich 1905 der Ansicht, dass man von dem Aether in der Physik überhaupt nicht sprechen dürfe. Dieses Urteil war aber zu radikal, wie wir bei den folgenden Überlegungen über die allgemeine Relativitätstheorie sehen werden. Es bleibt vielmehr nach wie vor erlaubt, ein raumfüllendes Medium anzunehmen als desser Zustände man die elektromagnetischen Felder (und wohl auch die Materie) ansehen kann. Aber es ist nicht gestattet, dieses Medium in Analogie zu der ponderabeln Materie in jedem Punkte einen Bewegungszustand zuzuschreiben. Dieser Aether darf nicht als aus Teilchen bestehend gedacht werden, deren Identität in der Zeit verfolgt werden könnte." A. Einstein, "Grundgedanken und Methoden der Relativitätstheorie in ihrer Entwicklung dargestellt"(*Morgan Manuscript*), *EA*, 2070, paragraph. 13.

**d7** "(22) *Allgemeine Relativitätstheorie und Aether.* Die Einfügung der im Rahmen der speziellen Relativitätstheorie bereits bekannten Naturgesetze in dem weiteren Rahmen der allgemeinen Relativitätstheorie hat keine Schwierigkeit. Die mathematischen Methoden lagen fertig vor in dem auf den Gauss-Riemann'schen Forschungen gegründeten "absoluten Differentialkalkül," die insbesondere von Ricci und Levi Civita ausgebaut worden war. Es handelt sich um einen einfachen Akt der Verallgemeinerung der Gleichungen vom Spezialfalle konstanten $g_{\mu\nu}$ zu dem Falle zeiträumlich veränderlicher $g_{\mu\nu}$. In allen so verallgemeinerten Gesetzen spielen die Gravitationspotentiale $g_{\mu\nu}$ eine Rolle, welche—kurz gesagt—die physikalischen Eigenschaften des leeren Raumes ausdrücken.

"Abermals erscheint also der "leere" Raum mit physikalischen Eigenschaften begabt, also nicht mehr als physikalisch leer, wie es nach der speziellen Relativitätstheorie schien. Man kann also sagen, dass der Aether in der allgemeinen Relativitätstheorie neu auferstanden ist, wenn auch in sublimierter Gestalt. Der Aether der allgemeinen Relativitätstheorie unterscheidet sich von dem der alten Optik darin, dass er kein Stoff im Sinne der Mechanik ist. Nicht einmal der Begriff der Bewegung kann auf ihn Anwendung finden. Er ist ferner keineswegs homogen und sein Zustand hat nicht selbständige Existenz sondern hängt ab von der feld-erzeugenden Materie. Da die metrischen Tatsachen von den "eigentlich" physikalischen in der neuen Theorie nicht mehr zu trennen sind, fliessen die Begriffe "Raum" und "Aether" zusammen. Da die Eigenschaften des Raumes als durch die Materie bedingt erscheinen, so ist nach der neuen Theorie der Raum nicht mehr ein Vorbedingung für die Materie; die Theorie vom Raume (Geometrie) und von der Zeit lässt sich nicht mehr der eigentlichen Physik voranstellen und unabhängig von Mechanik und Gravitation darlegen." A. Einstein, "Grundgedanken und Methoden der Relativitätstheorie in ihrer Entwicklung dargestellt" (*Morgan Manuscript*), EA, 2070, paragraph. 22.

**d8**    "Es trat plötzlich eine Organisation auf, deren alleiniger Zweck die Bekämpfung Einsteins und seiner Lehren war. Ihr Führer war ein gewisser Paul Weyland, über dessen Vergangenheit, Vorbildung und Beschäftigung niemand etwas wusste. Die Organisation verfügte über grosse Geldmittel unbekannter Herkunft. Sie bot verhältnismässig hohe Honorare für Leute, die gegen Einstein schreiben oder in Versammlungen gegen ihn sprechen wollten. Sie veranstaltete Versammlungen in dem grössten Konzertsaal Berlins und kündingte diese Versammlungen in Riesenplakaten an, wie es sonst nur bei Vorführungen der grössten Virtuosen üblich war." Ph. Frank, *Einstein. Sein Leben und seine Zeit* (Braunschweig/Wiesbaden: F. Vieweg u. Sohn, 1979), pp. 269-270.

**d9**    (**Original in Gothic script**): "Dass Einstein den Aether durch ein Dekret abschaffte, ihn aber durch einen anderen Begriff mit gleichen Funktionen wieder einführte, sei hier nur, um mit Einstein selbst zu reden, der "Drolligkeit" halber erwähnt." P. Weyland, "Einsteins Relativitätstheorie—eine wissenschaftlische Massensuggestion," *Täglische Rundschau*, Friday, 6/8/1920, evening edition.

**d10**    "Ich bin mir sehr wohl des Umstandes bewusst, dass die beiden Sprecher einer Antwort aus meiner Feder unwürdig sind; denn ich habe guten Grund zu glauben, dass andere Motiven als das Streben nach Wahrheit diesem Unternehmen zugrunde liegen. (Wäre ich Deutschnationaler mit oder ohne Hakenkreuz statt Jude freiheitlicher, internationaler Gesinnung, so ...). Ich antworte nur deshalb, weil dies von wohlwollender Seite wiederholt gewünscht worden ist, damit meine Auffassung bekannt werde." A. Einstein, "Meine Antwort über die antirelativitätstheoretische G.m.b.H.," *Berliner Tageblatt und Handelszeitung*, 27/8/1920, Nr. 402, pp. 1-2.

**d11**    "Als ausgesprochenen Gegner der Relativitätstheorie wüsste ich unter den Physikern von internationaler Bedeutung nur Lenard zu nennen. Ich bewundere Lenard als Meister der Experimentalphysik; in der theoretischen Physik aber hat er noch nichts geleistet, und seine Einwände gegen die allgemeine Relativitätstheorie sind von solcher Oberflächigkeit, dass ich es bis jetzt nicht für nötig erachtet habe, ausführlich auf dieselben zu antworten." A. Einstein, "Meine Antwort über die antirelativitätstheoretische G.m.b.H.," *Berliner Tageblatt und Handelszeitung*, 27/8/1920, n. 402, pp. 1-2.

d12  "Herr Gehrcke behauptet, das die Relativitätstheorie zum Solipsismus führe, eine Behauptung, die jeder Kenner als Witz begrüssen wird." A. Einstein, "Meine Antwort über die antirelativitätstheoretische G.m.b.H.," *Berliner Tageblatt und Handelszeitung*, 27/8/1920, Nr. 402, pp. 1-2.

d13  "Meine Stellung zur Ätherfrage werde ich ausführlich darlegen, sobald sich mir dazu Gelegenheit bietet." A. Einstein, Letter to H.A. Lorentz, 15/11/1919, *EA*, 16-494.

d14  "Die von Ihnen erwähnte Antrittsvorlesung will ich über Äther halten. Es ist eine schöne Gelegenheit, die von Ihnen angeregte Klarstellung vorzunehmen." A. Einstein, Letter to H.A. Lorentz, 12/1/1920, *EA*, 16-498.

d15  "Meine Antrittsvorlesung will ich über "Aether und Relativitätstheorie" halten, weil Lorentz schon bei meinem Besuch in Leiden wünschte, dass ich zu dieser Frage gelegentlich öffentlich Stellung nehme." A. Einstein, Letter to P. Ehrenfest, 12/1/1920, *EA*, 9-463.

d16  "Ich schreibe nun rüstig an der Antrittsvorlesung über den Äther, die natürlich nichts anderes sein kann als ein mehr oder weniger gefärbter Rückblick auf die Entwicklung unserer Meinungen von den physikalischen Eigenschaften des Raumes. Ich hoffe, dass wir in diesen fundamentalen Dingen nicht wesentlich verschiedener Meinung sind." A. Einstein, Letter to H.A. Lorentz, 18/3/1920, *EA*, 16-506.

d17  "Die Metrik ("der Zustand des Feldäthers") bestimmt eindeutig den affinen Zusammenhang (das "Gravitationsfeld")." H. Weyl, "Elektrizität und Gravitation," *PhZ*, 21 (1920), p. 649.

d18  "**Lenard**: Ich habe meine Meinung in der Druckschrift "Über Relativitätsprinzip, Äther, Gravitation" zum Ausdruck gebracht, dass der Äther in gewissen Beziehungen versagt hat, weil man ihn noch nicht in der rechten Weise behandelt hat. Das Relativitätsprinzip arbeitet mit einem nichteuklidischen Raum, der von Stelle zu Stelle und zeitlich nacheinander verschiedene Eigenschaften annimmt; dann kann nun eben in dem Raum ein Etwas sein, dessen Zustände diese verschiedenen Eigenschaften bedingen, und dieses Etwas ist eben der Äther. Ich sehe die Nützlichkeit des Relativitätsprinzips ein, solange es nur auf Gravitationskräfte angewandt wird. Für nicht massenproportionale Kräfte halte ich es für ungültig.

"**Einstein**: Es liegt in der Natur der Sache, dass von einer Gültigkeit des Relativitätsprinzip nur dann gesprochen werden kann, wenn es bezüglich aller Naturgesetze gilt.

"**Lenard**: Nur wenn man geeignete Felder hinzudichtet. Ich meine, das Relativitätsprinzip kann auch nur über Gravitation neue Aussagen machen, weil die im Falle der nichtmassenproportionalen Kräfte hinzugenommenen Gravitationsfelder gar keinen neuen Gesichtspunkt hinzufügen, als nur eben den, das Prinzip gültig erscheinen zu lassen. Auch macht die Gleichwertigkeit aller Bezugssysteme dem Prinzip Schwierigkeiten.

"**Einstein**: Es gibt kein durch seine Einfachheit prinzipiell bevorzugtes Koordinatensystem; deshalb gibt es auch keine Methode, um zwischen 'wirklichen' und 'nichtwirklichen' Gravitationsfeldern zu unterscheiden. Meine zweite Frage lautet: Was sagt das Relativitätsprinzip zu dem unerlaubten Gedankenexperiment, welches darin besteht, dass man z. B. die Erde ruhen und die übrige Welt um die Erdachse sich drehen lässt, wobei Überlicht- geschwindigkeiten aufheben?

"Der erste Satz ist keine Behauptung, sondern eine neuartige Definition für den Begriff 'Äther'.

"Ein Gedankenexperiment ist ein prinzipiell, wenn auch nicht faktisch ausführbares Experiment. Es dient dazu, wirklische Erfahrungen übersichtlich zusammenzufassen, um aus ihnen theoretische Folgerungen zu ziehen. Unerlaubt ist ein Gedankenexperiment nur dann, wenn eine Realisierung prinzipiell unmöglich ist.

"**Lenard**: Ich glaube zusammenzufassen zu können: **1**. Dass man doch besser unterlässt, die 'Abschaffung des Äthers' zu verkünden. **2**. Dass ich die Einschränkung des Relativitätsprinzip zu einem Gravitationsprinzip immer noch für angezeigt halte, und **3**., dass die Überlichtgeschwindigkeiten dem Relativitätsprinzip doch eine Schwierigkeit zu bereiten scheinen; denn sie heben bei der Relation jedes beliebigen Körpers auf, sobald man dieselbe nicht diesem, sondern der Gesamtwelt zuschreiben will, was aber das Relativitätsprinzip in seiner einfachsten und bisherigen Form als gleichwertig zulässt." "Allgemeine Diskussion über Relativitätstheorie," *PhZ*, 21 (1920), pp. 666-667.

**d19** **(Original in Gothic script):** "Die 'Abschaffung des Äthers' wurde in Nauheim in grosser Eröffnungssitzung wieder als Resultat verkündet (zur früheren Verkündung in Saltzburg, von Herrn Einstein selbst, siehe das Zitat in Note (7, S. 27), Man hat nicht dazu gelacht. Ich weiss nicht, ob es anders gewesen wäre, wenn die Abschaffung der Luft verkündet worden wäre" Ph. Lenard, "Zusatz betreffend die Nauheimer Diskussion über das Relativitätsprinzip, Äther, Gravitation," in: *Über Relativitätsprinzip, Äther, Gravitation* (Leipzig: Verlag von S. Hirzel 1921), p. 37, footnote 1.

**d20** "Nach Einstein ist die metrische Struktur des Äthers von der Art, wie sie Riemann annimmt." H. Weyl, "Die Relativitätstheorie auf der Naturforscherversammlung in Bad Nauheim," *Jahresbericht der Deutschen Mathematikvereinigung*, 31 (1922), pp. 51-63. The sentence is on p. 52.

**d21** "Das ist erstens die *Existenz des Äthers*. Lenard meint, Einstein habe, bei Aufstellung der speziellen Relativitätstheorie, allzu voreilig die Abschaffung des

Äthers verkündet. In der Tat kann er ja darauf hinweisen, dass Einstein heute wieder in der allgemeinen Relativitätstheorie von einem Äther spricht. Man darf sich doch aber durch das gleichlautende Wort nicht über die Verschiedenheit der Sache täuschen lassen! Der alte Äther der Lichttheorie war ein *substantielles* Medium, ein dreidimensionales Kontinuum, von welchem sich jede Stelle *P* in jedem Augenblick in einem bestimmten Raumpunkt *p* (oder an einer bestimmten Weltstelle) befindet; die Wiedererkennbarkeit derselben Ätherstelle zu verschiedenen Zeiten ist dabei das Wesentliche. Durch diesen Äther löst sich die vierdimensionale Welt auf in ein dreifach unendliches Kontinuum von eindimensionalen Weltlinien; infolgedessen gestattet er, *Ruhe* und *Bewegung* absolut voneinander zu unterscheiden. *In diesem Sinne*, etwas anderes hat Einstein nicht behauptet, ist der Äther durch die spezielle Relativitätstheorie abgeschafft; er wurde ersetzt durch die affingeometrische Struktur der Welt, welche nicht den Unterschied zwischen Ruhe und Bewegung festlegt, sondern die *gleichförmige Translation* von allen andern Bewegungen absondert. Der substantielle Äther war von seinen Erfindern als etwas Reales, den ponderablen Körpern Vergleichbares gedacht. In der Lorentzschen Elektrodynamik hatte er sich in eine rein geometrische, d. h. ein für allemal feste, von der Materie nicht beeinflusste Struktur verwandelt. In Einsteins spezieller Relativitätstheorie trat an ihre Stelle eine andere, die affingeometrische Struktur. In der allgemeinen Relativitätstheorie endlich verwandelte sich die letztere, als 'affiner Zusammenhang' oder 'Führungsfeld', wieder zurück in ein mit der Materie in Wirkungszusammenhang stehendes Zustandsfeld von physikalischer Realität. Und darum hielt es Einstein für angezeigt, das alte Wort Äther für den vollständig gewandelten Begriff wieder einzuführen; ob das zweckmässig war oder nicht, ist weniger eine physikalische als eine philologische Frage.

"*Zweitens*: die *Überlichtgeschwindigkeit.* Lenard meint, die allgemeine Relativitätstheorie führe die Überlichtgeschwindigkeit wieder ein, da sie als Bezugssystem z. B. die rotierende Erde zulässt; in hinreichend grossen Entfernungen treten dabei Überlichtgeschwindigkeiten auf. Dies ist ein offenbares Missverständnis. Sind $x_1$, $x_2$, $x_3$ die in bezug auf die rotierende Erde gemessenen Raumkoordinaten, $x_0$ die zugehörige 'Zeit' (auf ihre präzise Definition kommt es jetzt nicht an), so werden die Koordinatenlinien $x_0$, auf denen bei konstanten $x_1$, $x_2$, $x_3$ nur $x_0$ variiert, nicht alle zeitartige Richtung haben, d. h. es wird in diesen Koordinaten nicht überall $g_{00} > 0$ sein. Nun behauptet Einstein allerdings, dass auch solche Koordinatensysteme zulässig sind; auch in solchen Koordinatensystemen gelten seine allgemein invarianten Gravitationsgesetze. Dagegen hält er durchaus daran fest, dass die *Weltlinie eines materiellen Körpers* stets zeitartige Richtung besitzt, dass an einem materiellen Körper (und als 'Signalgeschwindigkeit') keine Überlichtgeschwindigkeit auftreten kann. Ein Koordinatensystem von der oben angegebenen Art lässt sich infolgedessen nicht in seiner ganzen Ausdehnung durch einen 'Bezugsmollusken' wiedergeben, d. h. man kann sich kein materielles Medium denken, dessen einzelne Elemente die Koordinatenlinien $x_0$ jenes Koordinatensystems als

Weltlinien beschreiben." H. Weyl, "Die Relativitätstheorie auf der Naturforscherversammlung in Bad Nauheim," *Jahresbericht der Deutschen Mathematikervereinigung,* 31 (1922), pp. 51-63. The quoted passage is on pp. 59-60.

**d22** "Dass der Tragödie am Schluss das Satyrspiel nicht fehle, entwickelte Hr. Rudolph eine phantastische Äthertheorie mit 'Lücken' zwischen fliessenden Ätherwänden, Sternfäden usw., mit Hilfe deren er aus Nichts die Sonnenmasse auf eine beliebige Anzahl von Dezimalen genau bestimmte ..." H. Weyl, "Die Relativitätstheorie auf der Naturforscherversammlung in Bad Nauheim," *Jahresbericht der Deutschen Mathematikervereinigung,* 31 (1922), pp. 51-63. The sentence is on p. 63.

**d23** "Da die Relativitätstheorie es leider noch nicht zustande gebracht hat, die für die Sitzung zur Verfügung stehende absolute Zeit von 9 bis 1 Uhr zu verlängern, muss die Sitzung vertagt werden." Ph. Frank, *Einstein. Sein Leben und seine Zeit* (Braunschweig: Vieweg, 1979), p. 275.

**d24** "Die Ätherhypothese musste aber stets im Denken der Physiker eine Rolle spielen, wenn auch zunächst meist nur eine latente Rolle." A. Einstein, *Äther und Relativitätstheorie.* Rede gehalten am 5. Mai 1920 an der Reichs-Universität zu Leiden (Berlin: Springer, 1920), p. 4.

**d25** "Was die mechanische Natur des Lorentzschen Äthers anlangt, so kann man etwas scherzhaft von ihm sagen, dass Unbeweglichkeit die einzige mechanische Eigenschaft sei, die ihm H.A. Lorentz noch gelassen hat. Man kann hinzufügen, dass die ganze Änderung der Ätherauffassung, welche die spezielle Relativitätstheorie brachte, darin bestand, dass sie dem Äther seine letzte mechanische Qualität, nämlich die Unbeweglichkeit, wegnahm." A. Einstein, *Äther und Relativitätstheorie.* Rede gehalten am 5. Mai 1920 an der Reichs-Universität zu Leiden (Berlin: Springer, 1920), 1920, p.7

**d26** "Indessen lehrt ein genaueres Nachdenken, dass diese Leugnung des Äthers nicht notwendig durch das spezielle Relativitätsprinzip gefordert wird. Man kann die Existenz eines Äthers annehmen; nur muss man darauf verzichten, ihm einen bestimmten Bewegungszustand zuzuschreiben, d. h. man muss ihm durch Abstraktion das letzte mechanische Merkmal nehmen, welches ihm Lorentz noch gelassen hatte." A. Einstein, *Äther und Relativitätstheorie.* Rede gehalten am 5. Mai 1920 an der Reichs-Universität zu Leiden (Berlin: Springer, 1920), p. 9.

**d27** "Es lassen sich ausgedehnte physikalische Gegenstände denken, auf welche der Bewegungsbegriff keine Anwendung finden kann. Sie dürfen nicht als aus Teilchen bestehend gedacht werden, die sich einzeln durch die Zeit hindurch verfolgen lassen. In der Sprache Minkowskis drückt sich dies so aus: nicht jedes in der vierdimensionalen Welt ausgedehnte Gebilde lässt sich als aus Weltfäden zusammengesetzt auffassen. Das spezielle Relativitätsprinzip verbietet uns, den Äther als aus zeitlich verfolgbaren Teilchen bestehend anzunehmen, aber die Ätherhypothese an sich widerstreitet der speziellen Relativitätstheorie nicht. Nur muss man sich davor hüten, dem Äther einen Bewegungszustand zuzusprechen." A. Einstein, *Äther und Relativitätstheorie.* Rede gehalten am 5. Mai 1920 an der Reichs-Universität zu Leiden (Berlin: Springer, 1920), p. 10.

**d28** "Anderseits lässt sich aber zugunsten der Ätherhypothese ein wichtiges Argument anführen. Den Äther leugnen, bedeutet letzten Endes annehmen, dass dem leeren Raume keinerlei physikalische Eigenschaften zukommen." A. Einstein, *Äther und Relativitätstheorie*. Rede gehalten am 5. Mai 1920 an der Reichs-Universität zu Leiden (Berlin: Springer, 1920), p. 11.

**d29** "Dieser Ätherbegriff, auf den die Machsche Betrachtungsweise führt, unterscheidet sich aber wesentlich vom Ätherbegriff Newtons, Fresnels und H. A. Lorentz'. Dieser Machsche Äther bedingt nicht nur das Verhalten der trägen Massen, sondern wird in seinem Zustand auch bedingt durch die trägen Massen." A. Einstein, *Äther und Relativitätstheorie*. Rede gehalten am 5. Mai 1920 an der Reichs-Universität zu Leiden (Berlin: Springer, 1920), p. 12.

**d30** "Der Machsche Gedanke findet seine volle Entfaltung in dem Äther der allgemeinen Relativitätstheorie. Nach dieser Theorie sind die metrischen Eigenschaften des Raum-Zeit-Kontinuums in der Umgebung der einzelnen Raum-Zeitpunkte verschieden und mitbedingt durch die ausserhalb des betrachteten Gebietes vorhandene Materie. Diese raum-zeitliche Veränderlichkeit der Beziehungen von Massstäben und Uhren zueinander, bzw. die Erkenntnis, dass der 'leere Raum' in physikalischer Beziehung weder homogen noch isotrop sei, welche uns dazu zwingt, seinen Zustand durch zehn Funktionen, die Gravitationspotentiale $g_{\mu\nu}$ zu beschreiben, hat die Auffassung, dass der Raum physikalisch leer sei, wohl endgültig beseitigt. Damit ist aber auch der Ätherbegriff wieder zu einem Inhalt, der von dem des Äthers der mechanischen Undulationstheorie des Lichtes weit verschieden ist. Der Äther der allgemeinen Relativitätstheorie ist ein Medium, welches selbst aller mechanischen und kinematischen Eigenschaften bar ist, aber das mechanische (und elektromagnetische) Geschehen mitbestimmt." A. Einstein, *Äther und Relativitätstheorie*. Rede gehalten am 5. Mai 1920 an der Reichs-Universität zu Leiden (Berlin: Springer, 1920), p.12.

**d31** "Kein Raum und auch kein Teil des Raumes ohne Gravitationspotentiale; denn diese verleihen ihm seine metrischen Eigenschaften, ohne welche er überhaupt nicht gedacht werden kann. Die Existenz des Gravitationsfeldes ist an die Existenz des Raumes unmittelbar gebunden." A. Einstein, *Äther und Relativitätstheorie*. Rede gehalten am 5. Mai 1920 an der Reichs-Universität zu Leiden (Berlin: Springer, 1920), pp. 13-14.

**d32** "Dagegen kann ein Raumteil sehr wohl ohne elektromagnetisches Feld gedacht werden; das elektromagnetische Feld scheint also im Gegensatz zum Gravitationsfeld gewissermassen nur sekundär an den Äther gebunden zu sein, indem die formale Natur des elektromagnetischen Feldes durch die des Gravitationsäther noch gar nicht bestimmt ist. Es sieht nach dem heutigen Zustande der Theorie so aus, als beruhe das elektromagnetische Feld dem Gravitationsfeld gegenüber auf einem völlig neuen formalen Motiv, als hätte die Natur den Gravitationsäther statt mit Feldern eines ganz anderen Typus, z.B. mit Feldern eines skalaren Potentials, ausstatten können.

"Da nach unseren heutigen Auffassungen auch die Elementarteilchen der Materie ihrem Wesen nach nichts anderes sind als Verdichtungen des elektromagnetischen Feldes, so kennt unser heutiges Weltbild zwei begrifflich vollkommen voneinander getrennte, wenn auch kausal aneinander gebundene Realitäten, nämlich Gravitationsäther und elektromagnetisches Feld oder—wie man sie auch nennen könnte—Raum und Materie.

"Natürlich wäre es ein grosser Fortschritt, wenn es gelingen würde, das Gravitationsfeld und das elektromagnetische Feld zusammen als ein einheitliches Gebilde aufzufassen." A. Einstein, *Äther und Relativitätstheorie*: Rede gehalten am 5. Mai 1920 an der Reichs-Universität zu Leiden (Berlin: Springer, 1920), p.14.

**d33**  "Ein überaus geistvoller Versuch in dieser Richtung ist von dem Mathematiker H. Weyl gemacht worden; doch glaube ich nicht, dass seine Theorie der Wirklichkeit gegenüber standhalten wird." A. Einstein, *Äther und Relativitätstheorie*. Rede gehalten am 5. Mai 1920 an der Reichs-Universität zu Leiden (Berlin: Springer, 1920), pp. 14-15.

**d34**  "Zusammenfassend können wir sagen: Nach der allgemeinen Relativitätstheorie ist der Raum mit physikalischen Qualitäten ausgestattet; es existiert also in diesem Sinne ein Äther. Gemäss der allgemeinen Relativitätstheorie ist ein Raum ohne Äther undenkbar; denn in einem solchen gäbe es nicht nur keine Lichtfortpflanzung, sondern auch keine Existenzmöglichkeit von Massstäben und Uhren, also auch keine räumlich-zeitlichen Entfernungen im Sinne der Physik. Dieser Äther darf aber nicht mit der für ponderable Medien charakteristischen Eigenschaft ausgestattet gedacht werden, aus durch die Zeit verfolgbaren Teilen zu bestehen; der Bewegungsbegriff darf auf ihn nicht angewendet werden." A. Einstein, *Äther und Relativitätstheorie*. Rede gehalten am 5. Mai 1920 an der Reichs-Universität zu Leiden (Berlin: Springer, 1920), p. 15

**d36**  "Das Buch von Eddington ist wirklich ausserordentlich geistvoll. Aber die Tendenz, die Naturgesetze nur als Ordnungsschemata zur Scheidung der darunter fallenden und der nicht darunter fallenden Fälle aufzufassen, kann ich doch nicht billigen. Auch ist zu beanstanden, dass er die Relativitätstheorie doch zu sehr als *logisch* notwendig darstellt. Gott hätte sich auch dazu entschliessen können, statt der relativistischen Aethers einen absolut ruhenden zu erschaffen. Dies gilt besonders dann, wenn er den Aether in de Sitter'scher (wesentlicher) Unabhängigkeit von der Materie eingerichtet haben sollte, zu welchem Glauben Eddington doch neigt; da kommt doch dem Aether ebenfalls eine "absolute" Funktion zu. Es ist merkwürdig, dass in den meisten Köpfen für die Wertung dieses Sachverhaltes kein Organ ist." A. Einstein, Letter to A. Sommerfeld, Nov, 28, 1926, [in:] A. Einstein, A. Sommerfeld, *Briefwechsel*, A. Hermann, ed. (Basel-Stuttgart: Schwabe u. Co. Verlag, 1968), p. 109.

**d37**  "[...] dass das Feld (oder der Äther) zwei Charakteristika enthält, nämlich das Gravitationspotential (oder die Metrik) und die elektrische Kraft." A. S. Eddington, *Relativitätstheorie in mathematischer Behandlung* (Berlin: Verlag J. Springer, 1925), p. 334.

**d38** "Diese ganze Anschauung gewinnt ausserordentlisch eine Einheitlichkeit, wenn der Äther nicht aus zwei miteinander in keinem inneren Zusammenhang stehenden, im extensiven Medium der Aussenwelt herrschenden Feldern besteht, sondern lediglich das mit einer von der Materie abhängigen metrischen Struktur begabte extensive Medium selber ist; so lehrt es die von mir aufgestellte erweiterte Relativitätstheorie." H. Weyl, "Feld und Materie," *AdP*, 65 (1921), pp. 541-563. The quoted passage is on p. 559.

**d39** "Wenn hier vom Äther die Rede ist, so soll es sich natürlich nicht um den körperlichen Äther der mechanischen Undulationstheorie handeln, welcher den Gesetzen der Newtonschen Mechanik unterliegt, und dessen einzelnen Punkten eine Geschwindigkeit zugeteilt wird. Dies theoretische Gebilde hat nach meiner Überzeugung seit der Aufstellung der speziellen Relativitätstheorie seine Rolle endgültig zu Ende gespielt. Es handelt sich vielmehr allgemeiner um diejenigen als physikalisch-real gedachten Dinge, welche neben der aus elektrischen Elementarteilchen bestehenden ponderabeln Materie im Kausal-Nexus der Physik eine Rolle spielen. Man könnte statt von "Äther" also ebensogut von "physikalischen Qualitäten des Raumes" sprechen." A. Einstein, "Über den Äther," *VSNG*, **105**, 1924, pp. 85-93. English translation: "On the Ether," in: *The Philosophy of Vacuum*, S. Saunders and H. R. Brown, eds. (Oxford: Clarendon Press, 1991), pp. 13-19. The quoted passage is on p. 13.

**d40** "Wir wollen dies physikalisch Reale, welches neben den beobachtbaren ponderabeln Körpern in das Newtonsche Bewegungsgesetzt eingeht, als 'Äther der Mechanik' bezeichnen." A. Einstein, "Über den Äther," *VSNG*, **105**, 1924, pp. 85-93. English translation: "On the Ether," in: *The Philosophy of Vacuum*, S. Saunders and H. R. Brown, eds. (Oxford: Clarendon Press, 1991), pp. 13-19. The sentence is on p. 15.

**d41** "Die zitierten Zeilen zeigen, dass Mach die schwachen Seiten der klassischen Mechanik klar erkannt hat und nicht weit davon entfernt war eine allgemeine Relativitätstheorie zu fordern, und dies schon vor einem halben Jahrhundert !." A. Einstein, "Ernst Mach," *PhZ*, 17 (1916), p. 103.

**d43** "Auch nach der speziellen Relativitätstheorie war der Äther absolut, denn sein Einfluss auf Trägheit und Lichtausbreitung war als unabhängig gedacht von Physikalischen Einflüssen jeder Art. [...] Also wird die Körpergeometrie wie die Dynamik vom Äther mitbedingt." A. Einstein, "Über den Äther," *VSNG*, **105**, 1924, pp. 85-93. English translation: "On the ether," in: *The Philosophy of Vacuum*, S. Saunders and H. R. Brown, eds. (Oxford: Clarendon Press, 1991), pp. 13-19. The sentence is on p. 17.

**d44** "Der Äther der allgemeinen Relativitätstheorie unterscheidet sich also vom demjenigen der klassischen Mechanik bezw. der speziellen Relativitätstheorie dadurch, dass er nicht 'absolut', sondern in seinen örtlich variablen Eigenschaften durch die ponderable Materie bestimmt ist." A. Einstein, "Über den Äther," *VSNG*, **105**, 1924, pp. 85-93. English translation: "On the Ether," in: *The Philosophy of Vacuum*, S. Saunders and H. R. Brown, eds. (Oxford: Clarendon Press, 1991), pp. 13-19. The sentence is on p. 18.

**d45** "Nun könnte allerdings die Meinung vertreten werden, dass unter diesen Begriff alle Gegenstände der Physik fallen, weil nach der konsequenten Feldtheorie auch die ponderable Materie, bzw. als besondere 'Raum-Zustände' aufzulassen seien." A. Einstein, "Über den Äther," *VSNG*, **105**, 1924, pp. 85-93. English translation: "On the Ether," in: *The Philosophy of Vacuum*, S. Saunders and H. R. Brown, eds. (Oxford: Clarendon Press, 1991), pp. 13-19. The sentence is on p. 13.

**d46** "Aber selbst wenn diese Möglichkeiten zu wirklichen Theorien heranreifen, werden wir des Äthers, d.h. des mit physikalischen Eigenschaften ausgestatteten Kontinuums, in der theoretischen Physik nicht entbehren können; denn die allgemeine Relativitätstheorie, an deren grundsätzlichen Gesichtspunkten die Physiker wohl stets festhalten werden, schliesst eine unvermittelte Fernwirkung aus; jede Nahewirkungs-Theorie aber setzt kontinuierliche Felder voraus, also auch die Existenz eines 'Äthers'." A. Einstein, "Über den Äther," *VSNG*, **105**, 1924, pp. 85-93. English translation: "On the Ether," in: *The Philosophy of Vacuum*, S. Saunders and H. R. Brown, eds. (Oxford: Clarendon Press, 1991), pp. 13-19. The sentence is on p. 20.

**d47** (Original in Gothic script) "Ich bringe dem Leser in dieser aufgeregten Zeit diese kleine objektive, leidenschaftslose Betrachtung, weil ich der Meinung bin, dass man durch stille Hingabe an die ewigen Ziele, die allen Kulturmenschen gemeinsam sind, der politischen Befundung heute wirksamer dienen kann als durch politische Betrachtungen und Bekentnisse." A. Einstein, "Induktion und Deduktion in der Physik," in der Morgenausgabe des *Berliner Tageblatt und Handelszeitung*, 48 Jahrgang, Nr 617, 25/12/1919, Suppl. 4.

**d48** (Original in Gothic script) "Die einfachste Vorstellung die man sich von der Entstehung einer Erfahrungswissenschaft bilden kann ist die nach der induktiven Methode. Einzeltatsachen werden so gewählt und gruppiert, dass der gesetzmässige Zusammenhang zwischen denselben klar hervortritt. Durch Gruppierung dieser Gesetzmässigkeiten lassen sich wieder allgemeinere Gesetzmässigkeiten erzielen, bis ein mehr oder weniger einheitliches System zu der vorhandenen Menge der Einzeltatsachen geschaffen wäre von der Art, das der rückschauende Geist aus den so gewonnenen letzten Verallgemeinerungen auf umgekehrtem, rein gedanklischem Weise wieder zu den Einzeltatsachen gelangen könnte." A. Einstein, "Induktion und Deduktion in der Physik," in der Morgenausgabe des *Berliner Tageblatt und Handelszeitung*, 48 Jahrgang, Nr 617, 25/12/1919, Suppl. 4.

**d49** (Original in Gothic script) "Schon ein flüchtiger Blick auf die tatsächliche Entwicklung lehrt, dass die grossen Fortschritte wissenschaftlicher Erkenntnis nur zum kleinen Teil auf diese Weise entstanden sind. Wenn nämlich der Forscher ohne irgendwelche vorgefasste Meinung an die Dinge heranginge, wie sollte er aus der ungeheuren Fülle kompliziertester Erfahrung überhaupt Tatsachen herausgreifen können, die einfach genug sind, um gesetzmassige Zusammenhänge offenbar werden zu lassen? Galilei hätte niemals das Gesetz des freien Falles finden können ohne die vorgefasste Meinung dass die Verhältnisse, welche wir tatsächlich vorfinden, durch die Wirkungen des Luftwiderstandes kompliziert seien, dass man also Fälle ins Auge fassen müsse, bei denen dieser eine möglichst geringe Rolle spielt.

"Die wahrhaft grossen Fortschritte der Naturerkenntnis sind auf einem der Induktion fast diametral entgegengesetzen Wege entstanden. Intuitive Erfassung des Wesentlischen eines grossen Tatsachenkomplexes führt den Forscher zur Aufstellung eines hypothetischen Grundgestzes oder mehrerer solcher. Aus dem Grundgesetz (System der Axiome) zieht er auf rein logisch-deduktivem Wege möglichst vollständig die Folgerungen. Diese oft erst durch langwierige Entwicklungen und Rechnungen aus dem Grundgesetz abzuleitenden Folgerungen lassen sich dann mit den Erfahrungen vergleichen und liefern so ein Kriterium für die Berechtigung des angenommenen Grundgesetzes. Grundgesetz (Axiome) und Folgerungen zusammen bilden dass was man eine "Theorie" nennt. Jeder Kündige weiss dass die grössten Fortschritte der Naturerkenntnis, zum Beispiel Newtons Gravitationstheorie, die Thermodynamik, die kinetische Gastheorie, die moderne Elektrodynamik, usw., alle auf solchem Wege entstanden sind, und dass ihrer Grundlage jener prinzipiell hypothetische Charakter zukommt. Der Forscher geht also zwar stets von den Tatsachen aus, deren Verknüpfung das Ziel seiner Bemühungen bildet. Aber er gelangt nicht auf methodischem, induktivem Wege zu seinem Gedankensysteme, sondern er schmiegt sich den Tatsachen an durch intuitive Auswahl unter den denkbaren, auf Axiomen beruhenden Theorien." A. Einstein, "Induktion und Deduktion in der Physik," in der Morgenausgabe des *Berliner Tageblatt und Handelszeitung*, 48 Jahrgang, Nr 617, 25/12/1919, Suppl. 4.

d50 **(Original in Gothic script)** "Eine Theorie kann also wohl als unrichtig erkannt werden wenn in ihren Deduktionen ein logischer Fehler ist, oder als unzutreffend, wenn eine Tatsache mit einer ihrer Folgerungen nicht im Einklang ist. Niemals aber kann die Wahrheit einer Theorie erwiesen werden. Denn niemals weiss man das auch in Zukunft keine Erfahrung bekannt werden wird, die ihren Folgerungen widerspricht; und stets sind noch andere Gedankensysteme denkbar, welche imstande sind, dieselben gegebenen Tatsachen zu verknüpfen. Stehen zwei Theorien zur Verfügung welche beide mit dem gegebenen Tatsachenmaterial vereinbar sind, so gibt es kein anderes Kriterium für die Bevorzugung der einen oder der anderen als den intuitiven Bild des Forschers. So ist es zu verstehen dass scharfsinnige Forscher die Theorien und Tatsachen beherrschen, doch leidenschaftliche Anhänger gegensätzlicher Theorien sein können." A. Einstein, "Induktion und Deduktion in der Physik," in der Morgenausgabe des *Berliner Tageblatt und Handelszeitung*, 48 Jahrgang, Nr 617, 25/12/1919, Suppl. 4.

d51 "[...] hat nach der allgemeinen Relativitätstheorie auch der in diesem Sinne leere Raum physikalische Qualitäten, welche durch die Komponenten des Gravitationspotentials mathematisch charakterisiert sind." A. Einstein, "Dialog über Einwände gegen die Relativitättheorie," *Die Naturwissenschaften*, **6**, 1918, p. 702.

d52 "Abermals erscheint also der "leere' Raum mit physikalischen Eigenschaften begabt, also nicht mehr als physikalisch leer, wie es nach der speziellen

Relativitätstheorie schien. Man kann also sagen, dass der Aether in der allgemeinen Relativitätstheorie neu auferstanden ist, wenn auch in sublimierter Gestalt." A. Einstein, "Grundgedanken und Methoden der Relativitätstheorie in ihrer Entwicklung dargestelt" (*Morgan Manuscript*), *EA*, 2070, par. 22.

**d53** "Einstein: Es gibt kein durch seine Einfachheit prinzipiell bevorzugtes Koordinatensystem [...] Der erste Satz is keine Behauptung, sondern eine neuartige Definition für den Begriff 'Äther'." "Allgemeine Diskussion über Relativitätstheorie," *PhZ*, 21 (1920), p. 667

**d54** "Nach meiner Meinung bildet die allgemeine Relativitätstheorie nämlich nur dann ein befriedigendes System, wenn nach ihr die physikalischen Qualitäten des Raumes allein durch die Materie vollständig bestimmt werden. Es darf also kein $g_{\mu\nu}$-Feld, d.h. kein Raum-Zeit-Kontinuum, möglich sein ohne Materie, welche es erzeugt." A. Einstein,"Kritisches zu einer von Hrn. De Sitter gegebenen Lösung der Gravitationsgleichungen," *SPAW*, 1Tl. (1918), pp. 270-272.

**d55** "*Machsches Prinzip*: Das *G*-Feld ist restlos durch die Massen der Körper bestimmt. Da Masse und Energie nach den Ergebnissen der speziellen Relativitätstheorie das Gleiche sind und die Energie formal durch den symmetrischen Energietensor ($T_{\mu\nu}$) beschrieben wird, so besagt dies, dass das *G*-Feld durch den Energietensor der Materie bedingt und bestimmt sei." A. Einstein, "Prinzipielles zur allgemeinen Relativitätstheorie," *AdP*, 55 (1918), pp. 241-244.

**d56** "Die vorstehenden Überlegungen zeigen die Möglichkeit einer theoretischen Konstruktion der Materie aus Gravitationsfeld und elektromagnetischem Felde allein." A. Einstein, "Spielen Gravitationsfelder im Aufbau der materiellen Elementarteilchen eine wesentliche Rolle," *SPAW*, 1Tl. (1919), pp. 349-356.

**d57** "Es bleibt vielmehr nach wie vor erlaubt, ein raumerfüllendes Medium anzunehmen als dessen Zustände man die elektromagnetischen Felder (und wohl auch die Materie) ansehen kann." A. Einstein, "Grundgedanken und Methoden der Relativitätstheorie in ihrer Entwicklung dargestellt" (*Morgan Manuscript*), *EA*, 2070, par. 13.

# e: Chapter 4

**e1** "Nach unablässigem Suchen in den letzten zwei Jahren glaube ich nun die wahre Lösung gefunden zu haben. Ich teile sie im folgenden mit." A. Einstein, "Einheitliche Feldtheorie von Gravitation und Elektrizität," *SPAW* (*pmK*) (1925), p. 414.

**e2** "Es sollen hier die Resultate weiterer Überlegungen gegeben werden, deren Ergebnisse mir sehr für Kaluzas Ideen zu sprechen scheinen." A. Einstein, "Zu Kaluzas Theorie des Zusamenhanges von Gravitation und Elektrizität," *SPAW* (*pmK*) (1927), p. 23.

**e3** "Zusammenfassend kann man sagen, dass Kaluzas Gedanke im Rahmen der allgemeinen Relativitätstheorie eine rationelle Begründung der Maxwellschen elektromagnetischen Gleichungen liefert und diese mit den Gravitationsgleichungen zu einem formalen Ganzen vereinigt." A. Einstein, "Zu Kaluzas Theorie des Zusamenhanges von Gravitation und Elektrizität," *SPAW* (*pmK*) (1927), p. 30.

e4    "Seitdem entdeckte ich, dass diese Theorie—wenigstens in erste Näherung—die Feldgesetze der Gravitation und des Elektromagnetismus ganz einfach und natürlich ergibt. Es ist daher denkbar, dass diese Theorie die ursprüngliche Fassung der allgemeinen Relativitätstheorie verdrängen wird." A. Einstein, "Neue Möglichkeiten für eine einheitlichen Feldtheorie von Gravitation und Elektrizität," *SPAW*(*pmK*) (1928), p. 224.

e5    "Eine tiefere Untersuchung der Konsequenzen der Feldgleichungen [...] wird zu zeigen haben, ob die Riemann-Metrik in Verbindung mit dem Fernparallelismus wirklich eine adäquate Auffassung der physikalischen Qualitäten des Raumes liefert. Nach dieser Untersuchung ist es nicht unwahrscheinlich." A. Einstein, "Zur einheitlichen Feldtheorie," *SPAW*(*pmK*) (1929), p. 7.

e6    "Unterdessen fand ich, dass es möglich ist, das Problem in völlig befriedingender Weise unter Zugrundelegung eines Hamilton-Prinzips zu lösen." A. Einstein, "Einheitliche Feldtheorie und Hamiltonisches Prinzip," *SPAW*(*pmK*) (1929), p. 156.

e10   "Physikalischer Raum und Äther sind nur verschiedene Ausdrücke für ein und dieselbe Sache; Felder sind physikalische Zustände des Raumes. Denn wenn dem Äther kein besonderer Bewegungs-Zustand zukommt, so scheint kein Grund dafür vorzuliegen, ihn neben dem Raume als ein Wesen besondered Art einzuführen." A.Einstein, "Das Raum- Äther- und Feld-Problem in der Physik," in: A. Einstein, *Mein Weltbild* (Amsterdam: Querido, 1934), p. 237.

e11   "Das Reale wird aufgefasst als vierdimensionales Kontinuum mit einer einheitlichen Struktur bestimmter Art (Metrik und Richtung). Die Gesetze sind Differentialgleichungen, welchen die genannte Struktur, d.h. die als Gravitation und Elektromagnetismus in Erscheinung tretenden Felder genügen. Die materiellen Teilchen sind Stellen hoher Felddichte ohne Singularität.

"Zusammenfassend können wir symbolisch sagen: Der Raum, ans Licht gebracht durch das körperliche Objekt, zur physikalischen Realität erhoben durch Newton, hat in der letzen Jahrzehnten den Äther und die Zeit verschlungen und scheint in Begriffe zu sein, auch das Feld und die Korpuskeln zu verschlingen, so dass er als alleiniger Träger der Realität übrig bleibt." A. Einstein, "Raum, Äther und Feld in der Physik," *FPh*, 1 (1930), pp. 173-180. The quoted passage is on pp. 179-180. English translation: *Ibid.* pp. 180-184.

e13   "Begriffe und Begriffs-Systeme dienen immer dazu, in unsere Erlebnisse Ordnung und 'Sinn' zu bringen." A. Einstein, "Das Raum-, Feld- und Äther-Problem in der Physik," *Die Koralle*, 5 (1930), p. 486.

e14   "Logisch betrachtet stammen Begriffe nie aus der Erfahrung: d.h. sie sind aus ihr allein nicht ableitbar. Und doch kommt ihre Bildung in unserem Geiste nur im

Hinblick auf das sinnlich erlebte zustande, und es ist die Erklärung derartiger fundamentaler Begriffe in der Aufzeigung desjenigen Charakters unserer Sinnen-Erlebnisse zu suchen, der zu der Bildung des Begriffes geführt hat." A. Einstein, "Raum, Äther und Feld in der Physik," *FPh*, 1 (1930), pp. 173-180. The sentence is on p. 173. English translation: *Ibid.* pp. 180-184.

e15    "Von dem angedeuteten Gesichtspunkte aus betrachtet, scheint nun der Raumbegriff derjenige des körperlichen Gegenstandes voranzugehen. Ist diese Begriffsbildung erfolgt, so heben sich als besonders einfach solche Erlebnis-Komplexe ab, die wir als 'Lagerung körperlicher Objekte' begrifflich kennzeichnen. Es ist klar, dass die Lagerungsbeziehungen der Körper im gleichen Sinne real sind wie die Körper selbst." A. Einstein,"Das Raum-, Feld- und Äther-Problem in der Physik," *Die Koralle*, 5, 1930, pp. 486-487. The sentence is on p. 486.

e16    "Also: ohne Körperbegriff kein Begriff räumlicher Relation zwischen Körpern und ohne den Begriff der räumlichen Relation kein Raumbegriff." A. Einstein, "Raum, Äther und Feld in der Physik," *FPh*, 1 (1930), pp. 173-180. The sentence is on p. 173. English translation: *Ibid.* pp. 180-184.

e17    "Bei Betrachtung der Lagerungsbeziehungen der Körper gegeneinander empfindet es nämlich der menschliche Geist als das Einfachere, die Lagen aller Körper auf die eines einzigen zu beziehen, als die verwirrende Mannigfaltigkeit jedes Körpers gegen alle anderen geistig zu verwirklichen. Dieser eine Körper, der allgegenwärtig und für alle anderen durchdringlich sein müsste, und mit allen in Berührung zu sein, ist uns allerdings nicht sinnlich gegeben, aber wir fingieren ihn zur Bequemlichkeit unseres Denkens." e43: A. Einstein, "Raum, Äther und Feld in der Physik," *FPh*, 1 (1930), pp. 173-180. The quoted passage is on p. 173. English translation: *Ibid.* pp. 180-184.

e18    "Denn in der ältesten Geometrie, welche uns die Griechen geschenkt haben, beschränkt sich die Untersuchung völlig auf die Lagerungbeziehungen idealisierter körperlicher Objekte, die dort "Punkt," "Gerade," "Ebene" heissen. In den Begriffen "Kongruenz" und "Messen" zeigt sich deutlich der Hinweis auf die Lagerungbeziehungen körperlicher Objekte. Ein räumliches Kontinuum, kurz gesagt "der Raum," kommt in der euklidischen Geometrie überhaupt nicht vor, trotzdem dieser Begriff dem vorwissenschaftlichen Denken natürlich bereits völlig geläufig war.

"Die ausserordentliche Bedeutung der Geometrie der Griechen liegt darin, dass sie den—soviel wir wissen—ersten gelungenen Versuch darstellt, einen Komplex sinnlicher Erfahrung durch ein logisch-deduktives System begrifflich zu erfassen." A. Einstein, "Raum, Äther und Feld in der Physik," *FPh*, 1 (1930), pp. 173-180. The quoted passage is on p. 174. English translation: *Ibid.* pp. 180-184.

e19    "Das räumliche Kontinuum wurde als solches erst durch die Modernen in die Geometrie eingeführt, und zwar durch DESCARTES, den Begründer der analytischen Geometrie. [...] vertiefte er die Geometrie als Wissenschaft in entscheidender Weise. Denn von nun an waren die Gerade und die Ebene gegenüber anderen Linien und Flächen nicht mehr prinzipiell bevorzugt, sondern alle Linien bzw. Flächen erfuhren nun eine gleichartige Behandlung. An die Stelle des kompliziert gebauten Axiomensystems der euklidischen Geometrie trat ein einziges Axiom, welches in unserer heutigen Ausdrucksweise so heisst: Es gibt Koordinatensysteme, gegenüber welchen der Abstand $ds$ benachbarter Punkte $P$ und $Q$ sich aus den Koordinatendifferenzen $dx_1$, $dx_2$, $dx_3$ durch den pythagoreischen Satz, d.h. durch die Formel

$$ds^2 = dx_1^2 + dx_2^2 + dx_3^2$$

ausdrückt. Hieraus, d.h. aus der euklidischen Metrik, lassen sich alle Begriffe und Sätze der euklidischen Geometrie deduzieren." A. Einstein, "Raum, Äther und Feld in der Physik," *FPh*, 1 (1930), pp. 173-180. The quoted passage is on pp. 174-175. English translation: *Ibid.* pp. 180-184.

e20 "Nach dem bisher Gesagten haben zwar die räumlichen Beziehungen der Körper physikalische Realität, nicht aber der Raum selbst. Dieser aber gewinnt physikalische Realität in Newtons Mechanik. Nach dieser tritt nämlich im Bewegungsgesetzt als Fundamentalbegriff die Beschleunigung auf. Beschleunigung ist dabei ein Bewegungszustand gegenüber dem Raume, der auf den Begriff der relativen Lagerung allein nicht zurückgeführt werden kann. Metrik und Trägheit sind also gemäss der Newtonschen Physik die wesentlichsten Eigenschaften des Raumes." A. Einstein, "Das Raum- Feld- und Äther-Problem in der Physik," *Die Koralle*, 5 (1930), pp. 486-487.

e21 "Anfangs waltete das Bestreben vor, das Feld als mechanischen Zustand einer überall gegenwärtigen Materie, des Äthers, aufzufassen. Als sich diese Bestrebung nicht befriedigend durchführen liess, hielt man zwar am Äther als einem besonderen Stoffe fest, dessen Zustände das Feld ausmachen sollten, aber die mechanische Interpretation dieser Zustände wurde fallen gelassen." A. Einstein, "Das Raum- Feld- und Äther-Problem in der Physik," *Die Koralle*, 5 (1930), p. 487.

e22 "Gegen Ende des vorigen Jahrhunderts zeigte H.A. Lorentz, dass dem Äther gegenüber dem Raume keine fortschreitende Bewegung zugeschrieben werden dürfte, wenn man die elektromagnetischen Vorgänge quantitativ richtig darstellen wollte." A. Einstein, "Das Raum- Feld- und Äther-Problem in der Physik," *Die Koralle*, 5 (1930), p. 487.

e23 "Wie nahe lag es zu sagen: die Felder sind Zustände des Raumes; Raum und Äther sind ein und dasselbe. Dass man es nicht sagte, lag daran, dass man den Raum als Sitz der euklidischen Metrik und der Galilei-Newtonschen Trägheit für absolut, d.h. unbeeinflussbar hielt, für ein starres Gerippe der Welt, das sozusagen vor aller Physik da ist und nicht Träger veränderlicher Zustände sein kann." A. Einstein, "Das Raum- Feld- und Äther-Problem in der Physik," *Transactions of the 2ⁿᵈ World Power Conference*, 19 (1930), p. 3.

e24 "Die Trennung der Begriffe Raum und Äther wurde so gewissermassen von selbst aufgehoben, nachdem bereits die spezielle Relativitätstheorie dem Äther den letzten Rest von Stofflichkeit genommen hatte." A. Einstein, "Das Raum- Feld-

und Äther-Problem in der Physik," *Transactions of the 2nd World Power Conference*, 19 (1930), p. 4.

**e25** "Damit hatte der Raum seinen absoluten Charakter verloren. Er war variabler (gesetzmässiger) Zustände und Vorgänge fähig, so dass er selbst die Funktionen des Äthers übernehmen konnte und—was das Gravitationsfeld anlangte—auch wirklich übernahm. Dunkel blieb vorläufig nur noch die formale Deutung des elektromagnetischen Feldes, das durch eine bloss metrische Struktur des Raumes sich nicht deuten liess." A. Einstein, "Das Raum- Feld- und Äther-Problem in der Physik," *Die Koralle*, 5 (1930), p. 487.

**e26** "Es musste also danach gestrebt werden, eine Struktur von grösserem Formenreichtum zu finden, welche die Riemannsche metrische Struktur in sich begreift und zugleich geeignet ist, das elektromagnetische Feld mathematisch zu beschreiben." A. Einstein, "Das Raum- Feld- und Äther-Problem in der Physik," *Transactions of the 2nd World Power Conference*, 19 (1930), p. 5.

**e27** "[...] physikalisher Raum und Äther sind nur verschiedene Ausdrücke für ein und dieselbe Sache; Felder sind physikalische Zustände des Raumes." A. Einstein, "Das Raum- Äther- und Feld-Problem in der Physik," in: A. Einstein, *Mein Weltbild* (Amsterdam: Querido, 1934), p. 237.

**e30** "[...] wollen wir hier eine Theorie geben, von der wir glauben, dass sie, abgesehen vom Quantenproblem, eine völlig befriedigende definitive Lösung bedeutet." A. Einstein, W. Mayer, "Eintheitliche Theorie von Gravitation und Elektrizität," *SPAW (pmK)* (1931), p. 541.

**e31** "Eine befriedigende Feldtheorie muss aber nach unserer Überzeugung mit einer singularitätsfreien Beschreibung des Gesamtfeldes, also auch des Feldes im Innern der Korpuskeln, auskommen. Deshalb stellten wir uns die Frage, ob die von uns betrachtete Raumstruktur nicht eine Verallgemeinerung zulasse, die zu elektromagnetischen Gleichungen mit nicht verschwindender elektrischer Dichte führen. Im folgenden soll gezeigt werden, dass es eine ganz natürliche derartige Verallgemeinerung gibt, welche zur Aufstellung eines kompatibeln Systems von Feldgleichungen Veranlassung gibt. Die Frage der Eignung dieses Gleichungssystems zur Beschreibung der Wirklichkeit soll hier noch nicht behandelt werden." A. Einstein, W. Mayer, "Eintheitliche Theorie von Gravitation und Elektrizität," *SPAW (pmK)* (1932), p. 130.

**e32** "Zwar habe ich, wie aus Vorstehendem hervorgeht, in dieser Schrift nicht eine Kritik der 'Relativitätstheorie' beabsichtigt—der Naturforscher sollte auch nur Besseres zu tun haben, als 'Theorie' zu kritisieren -, sondern es wird hier in der Hauptsache eine neue Auffassungsweise der Äthervorgänge mitgeteilt, durch welche die Relativitätstheorie überflüssig wird." Ph. Lenard, "Mahnwort zu Deutsche Naturforscher," in: Ph. Lenard, *Über Äther und Uräther*, 2. verm. Aufl. (Leipzig: Verlag von S. Hirzel), pp. 5 -10. The quoted passage is on p. 5.

**e33** "Ist es denn richtig, einen Hypothesenhaufen—und mag er mathematisch noch so schön aufgebaut sein—überhaupt vorzeitig 'Theorie' zu nennen? Schon die Ankündigung des Namens 'Relativitätstheorie' ist nach gegenwärtigem Stand der

Dinge Trug." Ph. Lenard, "Mahnwort zu Deutsche Naturforscher," in: Ph. Lenard, *Über Äther und Uräther*, 2. verm. Aufl. (Leipzig: Verlag von S. Hirzel), pp. 5 -10. The quoted passage is on p. 5.

**e34** "Hat nicht Herr Einstein den erst von ihm mit so grossem Nachdruck als nicht vorhanden hingestellten Äther dann später zum Gegenstand eines Vortrags gemacht, worin sein Vorhandensein und seine etwaigen Eigenschaften diskutiert werden? Ist der Äther diskutabel, so ist es auch die absolute Bewegung, und ist es diese, so ist es auch von Grund aus die Richtigkeit und aller Wert einer Relativitätstheorie; merkt man das nicht?" Ph. Lenard, "Mahnwort zu Deutsche Naturforscher," in: Ph. Lenard, *Über Äther und Uräther*, 2. verm. Aufl. (Leipzig: Verlag von S. Hirzel), pp. 5 -10. The passage is on pp. 5-6.

**e35** "Wo ist da deutsche Gediegenheit und Gründlichkeit geblieben?" Ph. Lenard, "Mahnwort zu Deutsche Naturforscher," in: Ph. Lenard, *Über Äther und Uräther*, 2. verm. Aufl. (Leipzig: Verlag von S. Hirzel), pp. 5 -10. The sentence is on p. 6.

**e36** "Wäre Einstein mit seiner Theorie doch von Anfang unter die Mathematiker und Philosophen gegangen! Die deutsche Physik wäre dann vielleicht von dem lähmenden Gift des Gedankens verschont geblieben, man könne aus geistreichen Fiktionen ("Gedankenexperimenten") mit Hilfe mathematischer Operationen physikalische Erkenntnisse oder, wie es in der Regel heisst, das "Weltbild" gewinnen." J. Starck, *Die gegenwärtige Krisis in der deutschen Physik* (Leipzig: Verlag von Johann A. Barth, 1922), p. 9.

**e37** "Die Übertreibung ins Abstrakte und Formale, die Beschränkung auf das intellektuelle Spiel mit mathematischen Definitionen und Formeln kommt in der Einsteinschen Relativitätstheorie vor allem in der absichtlichen Ignorierung des Äthers zum Ausdruck. Gewiss kann man physikalische Beziehungen zwischen materiellen Körpern in mathematischen Formeln unter Absehen vom Äther zwischen ihnen darstellen. Wird aber damit die Tatsache der Existenz des Äthers aus der Welt geschafft? In einer der Ansprachen auf der Nauheimer Naturforschungversammlung wurde es von einem Nichtphysiker als eine naturwissenschaftliche Grosstat Einsteins gefeiert, dass er den Äther abgeschafft habe. Soll man lachen über diese Wertschätzung einer vermeintlichen Grossleistung Einsteins, oder soll man empört sein über die von seinen Fiktionen angerichtete Verwüstung. Nein, die gefeierte Abschaffung des Äthers durch Einstein ist nicht eine Grosstat, sondern der Versuch zu einem verheerenden Rückschritt in der physikalischen Wissenschaft. Die Einführung des Ätherbegriffes in die Optik und die Elektrodynamik, das anschauliche Denken mit ihm hat sich in der Physik als ausserordentlich fruchtbar erwiesen; der Äther ist durch die physikalische Forschung eines Jahrhunderts aus einer Hypothese zu einer Tatsache geworden. Eine Physik ohne den Äther ist keine Physik. Einstein ist wohl selbst ob seiner Grosstat der Abschaffung des Äthers bange geworden; denn in neuerer Zeit scheint er in einem Vortrag den Äther wieder einführen zu wollen, freilich ist es nicht der alte abgeschaffte Äther, sondern eine Art Einsteinscher Relativitätsäther." J. Starck, *Die gegenwärtige Krisis in der deutschen Physik* (Leipzig: Verlag von Johann A. Barth, 1922), pp. 11-12.

**e38** "Hasenöhrl hat zuerst nachgewiesen, dass Energie Masse (Trägheit) besitzt" Ph. Lenard, *Über Energie und Gravitation* (Berlin-Leipzig: W. de Gruyter und Co., 1929), p. 3

**e41** "Zdawałem sobie również sprawę z tego, że jeśli książka ta ma mieć wartość historyczną, muszę pozostać w cieniu i pozwolić Einsteinowi wyrazić własne myśli" L. Infeld, *Szkice z przeszłości* (Warszawa: Państwowy Instytut Wydawniczy, 1964), pp. 54-55.

**e42** "Alle unsere Bemühungen dem Äther Wirklichkeit zu geben, schlugen fehl. Der Äther offenbart weder seine mechanischen Eigenschaften noch seine absolute Bewegung. Nichts bleibt von ihm erhalten, mit Ausnahme der Eigenschaft, für die er erfunden war, d.h. seine Fähigkeit, elektromagnetische Wellen zu übermitteln. Unsere Versuche, die Eigenschaften des Äthers herauszufinden, führten zu unüberwindlichen Schwierigkeiten und Widersprüchen. Nach solchen schlechten Erfahrungen ist es das beste, den Äther überhaupt vollständig zu vergessen und zu versuchen, seinen Namen nicht mehr zu nennen. Wir wollen sagen: "Unser Raum besitzt die physikalische Eigenschaft, Wellen zu übertragen" und enthalten uns des Gebrauches eines Wortes, das wir uns zu vermeiden entschlossen haben. Das Auslassen eines Wortes aus unserem Wörterbuch ist natürlich kein Heilmittel. Unsere Schwierigkeit ist in der Tat viel zu tiefgehend, als dass wir sie auf diese Weise lösen könnten." A. Einstein and L. Infeld. *Die Physik als Abenteuer der Erkenntnis* (Leiden: A.W. Sijthoffs Witgeversmaatschappij N.V., 1949), p. 116.

**e43** "Wir mögen noch weiterhin das Wort Äther gebrauchen, jedoch nur, um dadurch die physikalischen Eigenschaften des Raums auszudrücken. Dieses Wort Äther hat in der Entwicklung der Wissenschaft viele Male seine Bedeutung geändert. Gegenwärtig bezeichnet es nicht mehr ein Medium, das irgendwie aus materiellen Partikeln aufgebaut ist. Seine Geschichte ist aber noch keineswegs beendet und wird durch die Relativitätstheorie fortgesetzt werden." A. Einstein and L. Infeld. *Die Physik als Abenteuer der Erkenntnis* (Leiden: A.W. Sijthoffs Witgeversmaatschappij N.V., 1949), pp. 99-100.

**e44** "Aus der Relativitätstheorie wissen wir, dass die Materie ungeheure Energiespeicher darstellt, und das Energie Materie bedeutet. Auf diese Weise können wir nicht qualitativ zwischen Materie und Feld unterscheiden, da die Unterscheidung zwischen Masse und Energie keine qualitative ist [...] Es hat keinen Sinn, Materie und Feld als zwei gänzlich verschiedene Qualitäten zu betrachten. Wir können uns keine bestimmte Fläche vorstellen, die das Feld von der Materie scharf trennt [...] Die Einteilung in Materie und Feld ist aber nach der Erkenntnis der Äquivalenz von Masse und Energie etwas Künstlisches und nicht klar Definiertes [...] Materie ist dort, wo die Konzentration der Energie gross ist, Feld, wo die Konzentration der Energie klein ist. Wenn aber das der Fall ist, dann ist Unterschied zwischen Materie und Feld eher ein quantitativer als ein qualitativer." A. Einstein and L. Infeld. *Die Physik als Abenteuer der Erkenntnis* (Leiden: A.W. Sijthoffs Witgeversmaatschappij N.V., 1949), pp. 162-164.

**e45** "Ich glaube nun, nach langem Tasten die natürlichste Form für diese Verallgemeinerung gefunden zu haben." A. Einstein, "Relativität und

Raumproblem," in: A. Einstein, *Über die spezielle und die allgemeine Relativitätstheorie, gemeinverständlich* (Berlin: Akademie Verlag, 1969), p. 126 . English translation: A. Einstein, "Relativity and the Problem of Space," in A. Einstein, *Ideas and Opinions*, New York, p. 376.

**e47** "Gemäss der allgemeinen Relativitätstheorie dagegen hat der Raum gegenüber dem "Raum-Erfüllenden," von den Koordinaten Abhängigen keine Sonderexistenz [...] einen leeren Raum, d.h. einen Raum ohne Feld, gibt es nicht." A. Einstein, "Relativität und Raumproblem," in: A. Einstein, *Über die spezielle und die allgemeine Relativitätstheorie, gemeinverständlich* (Berlin: Akademie Verlag, 1969), p. 125. English translation: A. Einstein, "Relativity and the Problem of Space," in A. Einstein, *Ideas and Opinions*, New York, Crown Publ., 5th printing, 1960, p. 375. In the English version there is one more sentence. It reads as follows: "Space-time does not claim existence on its own, but only as a structural quality of the field."

**e49** "Dieser starre vierdimensionale Raum der speziellen Relativitätstheorie ist gewissermassen ein vierdimensionales Analogon des H.A. Lorentzschen starren dreidimensionalen Äthers." A. Einstein, "Relativität und Raumproblem," in: A. Einstein, *Über die spezielle und die allgemeine Relativitätstheorie, gemeinverständlich*, Akademie Verlag, Berlin, 1969, p. 121. English translation: A. Einstein, "Relativity and the Problem of Space," in: A. Einstein, *Ideas and Opinions* (New York: Crown Publ., 5th printing, 1960), pp. 360-377.

**e50** "Man könnte statt von "Äther" also ebensogut von "physikalischen Qualitäten des Raumes" sprechen." A. Einstein, "Über den Äther," *VSNG*, 105 (1924), p. 85.

**e51** "Die Ätherhypothese musste aber stets im Denken der Physiker eine Rolle spielen, wenn auch zunächst meist nur eine latente Rolle." A. Einstein, *Äther und Relativitätstheorie* (Berlin: Verlag von Julius Springer, 1920), p. 4.

**e52** "Dieses wurde auf Wunsch von Prof. Einstein weggelassen, da "die dort geschilderte Theorie von mir längst verlassen worden und durch die Theorie des nichtsymmetrischen Feldes ersetzt worden ist, die in logisch-formaler Beziehung völlig befriedigt." A. Einstein, Letter to C. Seelig, 1953, in: A. Einstein, *Mein Weltbild*, hrsg. von C. Seelig, mit Anmerkungen des Herausgebers (pp. 174-200) (Frankfurt/m-Berlin-Wien: Verlag Ullstein GmbH, 1983), p.199.

## f: Chapter 5

**f1** "Ich sage Ihnen glatt heraus: Die Physik ist ein Versuch der begrifflichen Konstruktion eines Modells der *realen* Welt sowie von deren gesetzlicher Struktur." A. Einstein, *Letter to M. Schlick*, 18/11/1930, English transl. in: Don Howard, "Realism and Conventionalism in Einstein's Philosophy of Science: The Einstein-Schlick Correspondence," *Philosophia Naturalis* 21 (1984), H. 2-4, p. 628.

**f3** "Gemäss der allgemeinen Relativitätstheorie ist ein Raum ohne Äther undenkbar; denn in einem solchen gäbe es nicht nur keine Lichtfortpflanzung, sondern auch keine Existenzmöglichkeit von Massstäben und Uhren, also auch keine räumlich-zeitlichen Entfernungen im Sinne der Physik." A. Einstein, *Äther und Relativitätstheorie* (Berlin: Verlag von J. Springer, 1920), p. 15.

f4   "Der vierdimensionale Raum der speziellen Relativitätstheorie ist ebenso starr und absolut wie der Raum Newtons." A. Einstein, "Das Raum-, Äther- und Feld-Problem der Physik," in: A. Einstein, *Mein Weltbild* (Amsterdam: Querido, 1934), p. 238.

f5   "Die Welt der Ereignisse kann dynamisch beschrieben werden durch ein Bild, das sich mit der Zeit ändert, und das auf den Hintergrund des dreidimensionalen Raumes geworfen wird. Sie kann aber auch durch ein statisches Bild beschrieben werden, das auf den Hintergrund eines vierdimensionalen Raum-Zeit-Kontinuum projiziert wird. Vom Standpunkt der klassischen Physik sind die beiden Bilder, das dynamische und das statische äquivalent. Vom Standpunkt der Relativitätstheorie aber ist das statische Bild bequemer und objektiver.

"Selbst in der Relativitätstheorie können wir, wenn wir es vorziehen, das dynamische Bild gebrauchen. Wir müssen uns aber daran erinnern, dass die Aufspaltung in Raum und Zeit keine objektive Bedeutung hat, da die Zeit nicht mehr "absolut" ist. Auf den folgenden Seiten werden wir uns noch weiter der "dynamischen" und nicht der "statischen" Sprache bedienen, aber uns ihrer Beschränkungen dabei immer bewusst sein." A. Einstein and L. Infeld. *Die Physik als Abenteuer der Erkenntnis* (Leiden: A.W. Sijthoffs Witgeversmaatschappij N.V., 1949), p. 140.

f7   "Wenn eine kleinere Schachtel *s* sich im Inneren des Hohlraumes einer grösseren Schachtel *S* in relativer Ruhe befindet, so ist der Hohlraum von *s* ein Teil des Hohlraumes von *S*, und zu beiden Schachteln gehört derselbe sie beide enthaltende "Raum." Weniger einfach aber ist die Auffassung, wenn s gegenüber *S* in Bewegung ist. Dann ist man geneigt zu denken, *s* umschliesse stets denselben Raum, aber einen veränderlichen Teil des Raumes *S*. Man ist dann genötigt , jeder Schachtel ihren besonderen (nicht als begrenzt gedachten) Raum zuzuordnen und anzunehmen, dass diese beiden Räume gegeneinander bewegt seien.

"Bevor man auf diese Komplikation aufmerksam geworden ist, erscheint der Raum als ein begrenztes Medium (Behälter), in dem die körperlichen Objekte herumschwimmen. Nun aber muss man denken, dass es unendlich viele Räume gibt, die gegeneinander bewegt sind. Der Begriff Raum als ein unabhängig von den Dingen objektiv Existierendes gehört schon dem vorwissenschaftlichen Denken an, nicht aber die Idee von der Existenz einer unendlichen Zahl von gegeneinander bewegten Räumen." A. Einstein, "Relativität und Raumproblem," in: A. Einstein, *Über die spezielle und die allgemeine Relativitätstheorie, gemeinverständlich* (Berlin: Akademie Verlag, 1969), p. 110. English translation: A. Einstein, "Relativity and the Problem of Space," in: A. Einstein, *Ideas and Opinions* (New York: Crown Publ., 5th printing, 1960), p. 362.

f9   "Bei Betrachtung der Lagerungsbeziehungen der Körper gegeneinander empfindet es nämlich der menschliche Geist als das Einfachere, die Lagen aller Körper auf die eines einzigen zu beziehen [...]. Dieser *eine* Körper, der allgegenwärtig und für alle anderen durchdringlich sein müsste, um mit allen in Berührung zu sein, ist uns allerdings nicht sinnlich gegeben, aber wir fingieren ihn zur Bequemlichkeit unseres Denkens" A. Einstein, "Raum, Äther und Feld in der Physik," *FPh*, 1 (1930), pp. 173-180. The quoted passage is on p. 173. English translation: *Ibid.* pp. 180-184.

**f10** "Es ist klar, dass in der ausserwissenschaftlichen Begriffswelt der Begriff des Raumes als eines realen Dinges wohl vorhanden war. Die Mathematik Euklids aber kannte diesen Begriff als solchen nicht, sondern behalf sich ausschliesslich mit den Begriffen Objekt, Lagebeziehungen zwischen Objekten. Punkt, Ebene, Gerade, Strecke sind die idealisierten körperlichen Objekte. Alle Lagenbeziehungen werden auf solche der Berührung (Schneiden von Geraden, Ebenen, Liegen von Punkten auf Geraden, *etc.*) zurückgeführt. Der Raum als Kontinuum kommt in dem Begriffsystem überhaupt nicht vor. Dieser Begriff wurde erst durch Descartes eingeführt, indem er den Raum-Punkt durch seine Koordinaten beschrieb. Erst hier erscheinen die geometrischen Gebilde gewissermassen als Teile des unendlichen Raumes, der als dreidimensionales Kontinuum aufgefasst wird. (...)

"Insofern die Geometrie als die Lehre von den Gesetzmässigkeiten der gegenseitigen Lagerung praktisch starrer Körper aufgefasst wird, ist sie als der älteste Zweig der Physik anzusehen. Diese Lehre konnte—wie schon bemerkt wurde—ohne den Raumbegriff als solchen auskommen, indem sie mit den idealen Körpergebilden Punkt, Gerade, Ebene, Strecke auskommen konnte. Hingegen hatte die Newtonsche Physik das Raumganze im Sinne Descartes' unbedingt nötig. Die Dynamik kommt nämlich mit den Begriffen Massenpunkt, (zeitlich variable) Entfernung zwischen Massenpunkten nicht aus. In den Newtonschen Bewegungsgleichungen spielt nämlich der Begriff der Beschleunigung eine fundamentale Rolle, welche durch die zeitlich variablen Punkt-Abstände allein nicht definiert werden kann. Newtons Beschleunigung ist nur als Beschleunigung gegen das Raumganze zu denken bezw. zu definieren." A. Einstein, "Das Raum-, Äther- und Feld-Problem der Physik," in: A. Einstein, *Mein Weltbild* (Amsterdam: Querido, 1934), pp. 233-234.

**f11** "[...] physikalischer Raum und Äther sind nur verschiedene Ausdrücke für ein und dieselbe Sache." A. Einstein, "Das Raum-, Äther- und Feld-Problem der Physik," in: A. Einstein, *Mein Weltbild* (Amsterdam: Querido, 1934), p. 237.

**f12** "Die Emanzipation des Feldbegriffes von der Annahme der Setzung eines materiellen Trägers gehört zu den psychologisch interessantesten Vorgängen in der Entwicklung des physikalischen Denkens." A. Einstein, "Relativität und Raumproblem," in: A. Einstein, *Über die spezielle und die allgemeine Relativitätstheorie, gemeinverständlich* (Berlin: Akademie Verlag, 1969). English transl. in: A. Einstein, "Relativity and the Problem of Space," in A. Einstein, *Ideas and Opinions* (New York, Crown Publ., 5th printing, 1960), pp. 360-377. The sentence is on p. 368.

**f15** "Aber diese neue Äthertheorie würde das Relativitätsprinzip nicht mehr verletzen. Denn der Zustand dieses $g_{\mu\nu}$ = Äthers wäre nicht der eines starren Körpers von selbständigen Bewegungszustande. Sondern ein Bewegungszustand wäre eine

Funktion des Ortes, bestimmt durch die materiellen Vorgänge." A. Einstein, Letter to H.A. Lorentz, 17/6/1916, *EA*, 16-453.

f16 "Es bleibt vielmehr nach wie vor erlaubt, ein raumerfüllendes Medium anzunehmen [...]. Aber es ist nicht gestattet, diesem Medium in Analogie zu der ponderabeln Materie in jedem Punkte einen Bewegungszustand zuzuschreiben. Diesere Aether darf nicht als aus Teilchen bestehend gedacht werden, deren Identität in der Zeit verfolgt werden könnte. [...] Da die metrischen Tatsachen von den "eigentlich" physikalischen in der neuen Theorie nicht mehr zu trennen sind, fliessen die Begriffe "Raum" und "Aether" zusammen." A. Einstein,"Grundgedanken und Methoden der Relativitätstheorie in ihrer Entwicklung dargestelt"*(Morgan Manuscript)*, *EA*, 2070.

f17 "[...] physikalischer Raum und Äther sind nur verschiedene Ausdrücke für ein und dieselbe Sache; Felder sind physikalische Zustände des Raumes. Denn wenn dem Äther kein besonderer Bewegungs-Zustand zukommt, so scheint kein Grund dafür vorzuliegen, ihn neben dem Raume als ein Wesen besonderer Art einzuführen" A. Einstein, "Das Raum-, Äther- und Feld-Problem der Physik," in: A. Einstein, *Mein Weltbild* (Amsterdam: Querido, 1934), p. 237.

f18 "Es lassen sich ausgedehnte physikalische Gegenstände denken auf welche der Bewegungsbegriff keine Anwendung finden kann. Sie dürfen nicht als aus Teilchen bestehend gedacht werden, die sich einzeln durch die Zeit hindurch verfolgen lassen. In der Sprache Minkowskis drückt sich dies so aus: nicht jedes in der vierdimensionalen Welt ausgedehnte Gebilde lässt sich als aus Weltfäden zusammengesetzt auffassen." A. Einstein, *Äther und Relativitätstheorie*: Rede gehalten am 5. Mai 1920 an der Reichs-Universität zu Leiden (Berlin: Springer, 1920), p. 10.

f19 "Ein Raum vom Typus (1), ist im Sinne der allgemeinen Relativitätstheorie nicht etwa ein Raum ohne Feld, sondern ein Spezialfall des $g_{ik}$-Feldes." A. Einstein, "Relativität und Raumproblem," in: A. Einstein, *Über die spezielle und die allgemeine Relativitätstheorie, gemeinverständlich*, Akademie Verlag, Berlin, 1969. English translation: A. Einstein, "Relativity and the Problem of Space," in: A. Einstein, *Ideas and Opinions* (New York: Crown Publ., 5th printing, 1960), pp. 360-377. The statement is on p. 375.

f20 "Gemäss der Relativitätstheorie gibt es keinen wesentlichen Unterschied zwischen Masse und Energie. Energie besitzt Masse and Masse repräsentiert Energie. Anstelle von zwei Erhaltungssätzen haben wir nur einen, denjenigen der Masse-Energie." A. Einstein and L. Infeld. *Die Physik als Abenteuer der Erkenntnis* (Leiden: A.W. Sijthoffs Witgeversmaatschappij N.V., 1949), p. 132.

f21 "[...] physikalischer Raum und Äther sind nur verschiedene Ausdrücke für ein und dieselbe Sache; Felder sind physikalische Zustände des Raumes." A.Einstein, "Das Raum- Äther- und Feld-Problem in der Physik," in: A. Einstein, *Mein Weltbild* (Amsterdam: Querido, 1934), p. 237.

**f22** "[...] hier der Raum nicht als etwas Selbständiges, sondern nur als kontinuerliches Feld von 4 Dimensionen auftritt." A. Einstein, in: *Cinquant'anni di relatività (1905-1955)* (Firenze: Giunti Barbèra, 1955), p.XVI.

**f23** "Wir mögen noch weiterhin das Wort Äther gebrauchen, jedoch nur, um dadurch die physikalischen Eigenschaften des Raums auszudrücken. Dieses Wort Äther hat in der Entwicklung der Wissenschaft viele Male seine Bedeutung geändert. Gegenwärtig bezeichnet es nicht mehr ein Medium, das irgendwie aus materiellen Partikeln aufgebaut ist. Seine Geschichte ist aber noch keineswegs beendet und wird durch die Relativitätstheorie fortgesetzt werden." A. Einstein and L. Infeld. *Die Physik als Abenteuer der Erkenntnis* (Leiden: A.W. Sijthoffs Witgeversmaatschappij N.V., 1949), pp. 99 -100.

**f24** "Aber selbst wenn diese Möglichkeiten zu wirklichen Theorien heranreifen, werden wir des Äthers, d.h. des mit physikalischen Eigenschaften ausgestatteten Kontinuums, in der theoretischen Physik nicht entbehren können; denn die allgemeine Relativitätstheorie, an deren grundsätzlichen Gesichtspunkten die Physiker wohl stets festhalten werden, schliesst eine unvermittelte Fernwirkung aus; jede Nahewirkungs-Theorie aber setzt kontinuierliche Felder voraus, also auch die Existenz eines 'Äthers'." A. Einstein, "Über den Äther," *VSNG*, 105 (1924), pp. 85-93. English translation: "On the ether," in: *The Philosophy of Vacuum*, S. Saunders and H. R. Brown, eds. (Oxford: Clarendon Press, 1991), pp. 13-19. The sentence is on p. 20.

# Bibliography

Aiton, E. J. "Newton's Aether-Stream Hypothesis and the Inverse Square Law of Gravitation." *ASc*, 25 (1969), pp.225-260.

Arodź, H. "Albert Einstein a problem unifikacji oddziaływań fundamentalnych." (Albert Einstein and the problem of unification of the fundamental interactions) *PF*, 37 (1986), No. 4, pp. 297-309

Avenarius, R. *Kritik der reiner Erfahrung*. Leipzig: Fues [R. Reisland], 1888-1890.

Barone, M., and F. Selleri, eds. *Frontiers of Fundamental Physics*. New York-London-Boston: Plenum Publ. Co., 1994, pp. 193-201.

Born, M. "Physik und Relativität" in: M. Born, *Physik im Wandel meiner Zeit*. Vierte, erweiterte Auflage. Braunschweig: F. Vieweg und Sohn, 1966, pp. 185-201.

Cantor, G. N., and M.J.S. Hodge, eds. *Conceptions of Ether*. Cambridge: Cambridge University Press, 1981.

Davies, P.C.W., and J. R. Brown, eds. *The Ghost in the Atom*. Cambridge: Cambridge University Press, 1986, pp. 49-50.

Dingler, Hugo. "Kritische Bemerkungen zu Grundlagen der Relativitätstheorie." *PhZ*, 21 (1920), pp. 668-675.

Drude, P. *Physik des Äthers auf Elektromagnetischer Grundlage*. Stuttgart: Verlag von F. Enke, 1894.

Drude, P. *Die Theorie in der Physik*. Leipzig: Verlag von S. Hirzel, 1895.

Drude, P. *Lehrbuch der Optik*. Leipzig: Verlag von S. Hirzel, 1900.

Duffy, M. C., ed. *Physical Interpretations of Relativity*. London: Imperial College, Proceedings of conferences held in 1988, 1990, 1992, 1994, 1996, 1998

Eddington, A. S. *Space, Time and Gravitation. An Outline of the General Relativity Theory*. Cambridge: Cambridge University Press, 1920.

Eddington, A. S. *Relativitätstheorie in mathematischer Behandlung*. Berlin: Verlag von J. Springer, 1925.

Einstein, A. "Über die Untersuchung des Aetherzustandes im magnetischen Felde." [1894 or 1895], *PhB*, 27 (1971), pp. 390-391.

Einstein, A. "Letter to M. Marič, July 1899?" *EA* FK-53

Einstein, A. "Folgerungen aus den Kapillaritätserscheinungen." *AdP*, 4 (1901), pp. 513-523.

Einstein, A. "Letter to W. Ostwald, 19/3/1901." in: H.G. Körber, "Zur Biographie des jungen Albert Einstein." *Forschungen und Fortschritte*, 38 (1964), pp. 75-76.

Einstein, A. "Letter to M. Grossmann, 6?/9/1901." *EA* 11-485.

Einstein, A. reviews in: *Beiblätter zu den Annalen der Physik* No. 29 (1905), pp. 235-238, 240-242, 246-247, 623-624, 629, 634-636, 640-641, 950, 952-953, 1114-1115, 1152-1152-1153, 1158

Einstein, A. "Über einen die Erzeugung und Verwandlung des Lichtes betreffenden heuristischen Gesichtspunkt." *AdP*, 17 (1905), pp. 132-148.

Einstein, A. "Zur Elektrodynamik bewegter Körper." *AdP*, 17 (1905), pp. 891-921.

Einstein, A. "Ist die Trägheit eines Körpers von seinem Energieinhalt abhängig?" *AdP*, 18 (1905), pp. 639-641.

Einstein, A. *Eine neue Bestimmung der Moleküldimensionen.* Bern: Wyss, 1905; also in: *AdP*, 19 (1906), pp. 286-306.

Einstein, A. "Relativitätsprinzip und die aus demselben gezogenen Folgerungen." *JR*, 4 (1907), pp. 411-462; and *JR*, 5 (1907), pp. 98-99.

Einstein, A. "Zum gegenwärtigen Stande des Strahlungsproblems." *PhZ*, 10 (1909), pp. 185-193.

Einstein, A. "Letter to J. J. Laub, undated, about 1908." in: A. Kleinert, Ch. Schönbeck, "Ph. Lenard und A. Einstein. Ihr Briefwechsel und ihr Verhältnis vor der Nauheimer Diskussion." *Gesnerus*, 35 (1978), pp. 318-333 (see p. 320).

Einstein, A. "Entwicklung unserer Anschauungen über das Wesen und die Konstitution der Strahlung." *PhZ*, 10 (1909), p. 817-825.

Einstein, A. "Letter to J. J. Laub, 16/3/1910." in: A. Kleinert, Ch. Schönbeck, "Ph. Lenard und A. Einstein. Ihr Briefwechsel und ihr Verhältnis vor der Nauheimer Diskussion." *Gesnerus*, 35 (1978), pp. 318-333 (see p. 320).

Einstein, A. "Principe de relativité et ses conséquences dans la physique moderne." *ASPN*, 29 (1910), pp. 5-28 and 125-244.

Einstein, A. "Letter to J. J. Laub, undated." in: A. Kleinert, Ch. Schönbeck, "Ph. Lenard und A. Einstein. Ihr Briefwechsel und ihr Verhältnis vor der Nauheimer Diskussion." *Gesnerus*, 35 (1978), pp. 318-333 (see p. 322).

Einstein, A. "Relativitätstheorie." *VNGZ*, 56 (1911), pp. 1-14.

Einstein, A. "Einfluss der Schwerkraft auf die Ausbreitung des Lichtes." *AdP*, 35 (1911), pp.898-908.

Einstein, A. "Lichtgeschwindigkeit und Statik des Gravitationsfeldes." *AdP*, 38 (1912), pp. 355-369.

Einstein, A. "Zur Theorie des statisches Gravitationsfeldes." *AdP*, 38 (1912), pp. 443-458.

Einstein, A. "Gibt es eine Gravitationswirkung die der elektrodynamischen Induktionswirkung analog ist?" *VgM*, 44 (1912), pp. 37-40.

Einstein, A. Grossmann M., "Entwurf einer verallgemeinerten Relativitätstheorie und einer Theorie der Gravitation." *ZMPh*, 62 (1913), p. 225-261.

Einstein, A. "Physikalische Grundlagen einer Gravitationstheorie." *VNGZ*, 58 (1913), pp. 284-290.

Einstein, A. "Gravitationstheorie." *VSNG* (1913), 2. Tl., pp. 137-138.

Einstein, A. "Letter to E. Mach, undated." in: V.P. Vizgin, Ya.A. Smorodinskii. "From the equivalence principle to the equation of gravitation." *Sov. Phys. Usp.*, 22 (7), July 1979, p. 499.

Einstein, A. "Zum Relativitätsproblem." *Scientia*, 15 (1914), pp. 337-348.

Einstein, A. "Formale Grundlage der allgemeine Relativitäts Theorie." *SPAW* (1914), 2. Tl., pp. 1030-1085.

Einstein, A., and M. Grossmann. "Kovarianzeigenschaften der Feildgleichungem der auf die verallgemeinerte Relativitätstheorie gegründeten Gravitationstheorie." *ZMPh*, 63 (1914), pp. 215-225.

Einstein, A. "Vom Relativitätsprinzip." *Die Vossische Zeitung* Nr. 209: 1, 26/4/1914.

Einstein, A. "Antrittsrede." *SPAW*, 2. Tl. (1914), pp. 739-742.

Einstein, A. "Relativitätstheorie." in: *Die Physik*. unter Redaktion von E. Lecher, Leipzig: Teubner, 1915, pp.702-713.

Einstein, A. "Letter to P. Ehrenfest." quoted in: Klein, M.J. *Paul Ehrenfest. Vol. 1: The Making of Theoretical Physicist*. Amsterdam: North-Holland, 1972, p. 190; and in: F. Selleri. *Le grand débat de la théorie quantique*. Paris: Flammarion, 1986, p. 24.

Einstein, A. "Zur allgemeinen Relativitätstheorie." *SPAW* (1915), 2. Tl., pp. 778-786.

Einstein, A. "Zur allgemeinen Relativitätstheorie, Nachtrag." *SPAW* (1915), 2. Tl., pp. 799-801.

Einstein, A. "Erklärung der Perihelbewegung des Merkur aus der allgemeinen Relativitätstheorie." *SPAW* (1915), 2. Tl., pp. 831-839.

Einstein, A. "Feldgleichungen der Gravitation." *SPAW* (1915), 2. Tl., pp. 844-847.

Einstein, A. "Die Grundlage der allgemeinen Relativitätstheorie." *AdP*, 49 (1916), pp. 769-822.

Einstein, A. "Letter to H. A. Lorentz, 17/6/1916." *EA* 16-453.

Einstein, A. "Ernst Mach." *PhZ*, 17 (1916), pp. 101-104.

Einstein, A. [review] "H. Weyl, Raum, Zeit, Materie." *Die Naturwissemschaften* 6 (1918), p. 373.

Einstein, A. "Dialog über Einwände gegen die Relativitätstheorie." *Die Naturwissenschaften*, 6 (1918), pp. 697-702.

Einstein, A. "Bemerkung zu E. Gehrckes Notiz: Über den Äther." *VDPG*, 20 (1918), p. 261.

Einstein, A. "Motiv des Forschers." in: *Zu Max Plancks 60. Geburtstag: Ansprachen in der deutschen Physikalischen Geselschaft*. Karlsruhe: Müller, 1918, pp. 29-32.

Einstein, A. "Kritisches zu einer von Herrn de Sitter gegebenen Lösung der Gravitationsgleichungen." *SPAW*, 1. Tl. (1918), pp. 270-272.

Einstein, A. "Prinzipielles zur allgemeinen Relativitätstheorie." *AdP*, 55 (1918), pp. 241-244.

Einstein, A. "Spielen Gravitationsfelder im Aufbau der materiallen Elementarteilchen eine wesentliche Rolle?." *SPAW*, 1. Tl. (1919), pp. 349-356.

Einstein, A. "Letter to H.A. Lorentz, 15/11/1919." *EA* 16-494.

Einstein, A. "Induktion und Deduktion in der Physik." in: *Berliner Tageblatt und Handelszeitung*. Morgenausgabe, 48. Jahrgang, Nr 617, 25/12/1919, Suppl. 4.

Einstein, A. "Grundgedanken und Methoden der Relativitätstheorie in ihrer Entwicklung dargestellt." (*Morgan Manuscript*), *EA* 2070.

Einstein, A. "Meine Antwort über die antirelativitätstheoretische G.m.b.H." *Berliner Tageblatt und Handelszeitung*, 27 August 1920, Nr. 402, pp. 1-2.

Einstein, A., Ph. Lenard *et al.* "Allgemeine Diskussion über Relativitätstheorie." *PhZ*, 21 (1920), pp. 666-668.

Einstein, A. *Äther und Relativitätstheorie*. Berlin: Verlag von J. Springer, 1920.

Einstein, A. "Letter to H.A. Lorentz, 12/1/1920." *EA* 16-498.

Einstein, A. "Letter to P. Ehrenfest, 12 /1/1920." *EA* 9-463.

Einstein, A. "Letter to H.A. Lorentz, 18/3/1920." *EA* 16-506.

Einstein, A. "A brief outline of the development of the theory of relativity." *Nature*, 106 (1921), pp. 782-784.

Einstein, A. "Speech at Kyoto University, 14/12/1922." *NTM-Schriftenreihe für Geschichte der Naturwissenschaften, Technik und Medizin, Leipzig*, 20 (1983), pp. 25-28.

Einstein, A., "Speech at Kyoto University, December 14, 1922." *Physics Today*, 25 (1982), pp. 45-47.

"Einstein and the Philosophies of Kant and Mach (an interview with A. Einstein)" *Nature*, 112 (1923), p. 253.

Einstein, A. "Über den Äther." *VSNG*, 105 (1924), pp. 85-93.

Einstein, A. "Einheitliche Feldtheorie von Gravitation und Elektrizität." *SPAW (pmK)* (1925), pp.. 414-419.

Einstein, A. "Letter to A. Sommerfeld, 28/11/1926." in: A. Einstein and A. Sommerfeld. *Briefwechsel*. Basel-Stuttgart: Schwabe u. Co. Verlag, 1968, p. 109.

Einstein, A. "Zu Kaluzas theorie des Zusammenhanges von Gravitation und Elektrizität." *SPAW (pmK)* (1927), pp. 23-30.

Einstein, A. "Riemanngeometrie mit Aufrechterhaltung des Begriffes des Ferm-Parallelismus." *SPAW (pmK)* (1928), pp. 217-221.

Einstein, A. "Neue Möglichkeiten für eine einheitliche Feldtheorie von Gravitation und Elektrizität." *SPAW (pmK)* (1928), pp. 224-227.

Einstein, A. "Auf die Riemann-Metrik und Fern-Parallelismus gegründete einheitliche Feldtheorie." *MA*, 102 (1929), pp. 685-697.

Einstein, A. "Zur einheitlichen Feldtheorie." *SPAW (pmK)* (1929), pp. 2-7.

Einstein, A. "Einheitliche Feldtheorie und Hamiltonsches Prinzip." *SPAW (pmK)* (1929), pp. 156-159.

Einstein, A. "An interview with A. Einstein." *Daily Chronicle*, 26/1/1929; *Nature*, 123 (1929), p. 175.

Einstein, A. "Field Theories, Old and New." *New York Times*, 3/2/1929.

Einstein, A. "The New Field Theory." *Times*. London, 4/2/1929.

Einstein, A. "The Concept of Space" (prepared by H.T.H. Piaggio), *Nature*, 125 (1930), pp. 897-898.

Einstein, A. "The New Field Theory." *Observatory*, 52 (1930), pp. 82-87 and 114-118.

Einstein, A. "Raum, Äther und Feld in der Physik." *FPh*, 1 (1930), pp. 173-180.

Einstein, A. "Space, Ether and Field in Physics." *FPh*, 1 (1930), pp. 180-184.

Einstein, A. "Das Raum-, Feld- und Äther-Problem in der Physik." *Die Koralle*, 5 (1930), H. 11, pp. 486-487.

Einstein, A. "Address at the University of Nottingham." *Science*, 71 (1930), pp. 608-610 (translated by. I. H. Brose).

Einstein, A. "Das Raum-, Feld- und Äther-Problem in der Physik." *Transactions of the Second World Power Conference*, 19 (1930), pp. 1-5.

Einstein, A. "Die Kompatibilität der Feldgleichungen in der einheitlichen Feldtheorie." *SPAW (pmK)* (1930), pp.18-23.

Einstein, A., and W. Mayer. "Zwei strenge statische Lösungen der Feldgleichungen der einheitlichen Feldtheorie." *SPAW (pmK)* (1930), pp. 110-120.

Einstein, A. "Zur Theorie der Räume mit Riemann-Metrik und Fernparallelismus." *SPAW (pmK)* (1930), pp.401-402.

Einstein, A. "Letter to M. Schlick, 18/11/1930." in: Don Howard, "Realism and Conventionalism in Einstein's Philosophy of Science: The Einstein-Schlick Correspondence." *Philosophia Naturalis*, 21 (1984), H. 2-4, pp. 628.

Einstein A., and W. Mayer. "Systematische Untersuchung über kompatible Feldgleichungen welche in einen Riemannschen Raume mit Fernparallelismus gesetzt werden können." *SPAW (pmK)* (1931), pp. 257-265.

Einstein, A. "Theory of Relativity: Its Formal Content and its Present Problems (a short presentation." in: *Nature*, 127 (1931), pp. 765, 790, 826-827.

Einstein, A. "Gravitational and electrical fields (Preliminary report for the Josiah Macy, Jr. Foundation)" *Science*, 73 (1931), pp. 438-439.

Einstein, A., and W. Mayer. "Einheitliche Theorie von Gravitation und Elektrizität." *SPAW (pmK)* (1931), pp. 541-557.

Einstein, A., and W. Mayer. "Einheitliche Theorie von Gravitation und Elektrizität." 2. Abhandlung, *SPAW (pMK)* (1932), pp. 130-137.

Einstein, A. "Paul Ehrenfest *in memoriam* (1934)." in: A. Einstein, *Out of My Later Years*. New York: Philosophical Library, 1950.

Einstein, A. "Das Raum-, Äther- und Feld-Problem der Physik." in: A. Einstein, *Mein Weltbild*. Amsterdam: Querido, 1934, pp. 229-248.

Einstein, A. *Mein Weltbild*. Amsterdam: Querido, 1934.

Einstein, A. Rosen N., "The Particle Problem in the General Theory of Relativity." *PhR*, 48 (1935), pp. 73-77.

Einstein, A. "Physik und Realität." *JFI*, 221 (1936), pp. 313-347.

Einstein, A., L. Infeld., and B. Hoffmann. "Gravitational equations and the problems of motion." *AoM*, 39 (1938), pp. 65-100.

Einstein, A., and L. Infeld. *Die Physik als Abenteuer der Erkenntnis*. Leiden: A. W. Sijthoffs Witgeversmaatschappij N. V., 1938 (new edition., 1949); the English translation: *The Evolution of Physics*. New York: Simon and Schuster, 1938; (also Cambridge University Press, 1938).

Einstein, A., and P. Bergmann. "Generalisation of Kaluza's theory of electricity." *AoM*, 39 (1938), pp. 683-701.

Einstein, A., V. Bargmann and P. G. Bergmann. "Five-dimensional representation of gravitation and electricity." in: *Theodore von Karman Anniversary Volume*. Pasadena: California Institute of Technology, 1941, pp. 212-225.

Einstein, A., and V. Bargmann. "Bivector fields I." *AoM*, 45 (1944), pp. 1-14.

Einstein, A. "Bivector fields II." *AoM*, 45 (1944), pp. 15-23.

Einstein, A. "Generalisation of the relativistic theory of gravitation I." *AoM*, 46 (1945), pp. 578-584.

Einstein, A., and E.G. Strauss. "Generalisation of the relativistic theory of gravitation II." *AoM*, 47 (1946), pp. 731-741.

Einstein, A. "Autobiographisches (1946)." in: *Albert Einstein als Philosophe und Naturforscher*. P. A. Schilpp ed. Braunschweig/Wiesbaden: Friedr. Vieweg und Sohn, 1979, p. 1-35.

Einstein, A. *The Meaning of Relativity*. Princeton: Princeton Univ. Press, 1950.

Einstein, A. "Generalised Theory of Gravitation." As Appendix II to *The Meaning of Relativity*. Princeton: Princeton Univ. Press, 1950.

Einstein, A. *Out of My Later Years*. New York: Philosophical Library, Inc., 1950.

Einstein, A. "On the generalised theory of gravitation." *Scientific American* 182 (1950), pp. 13-17.

Einstein, A. "Letter to C. Seelig, 8/4/1952." in: G. Holton, *Thematische Analyse der Wissenschaft, Die Physik Einsteins und seiner Zeit*. Frankfurt am Main: Suhrkamp Verlag, 1981, pp. 203-254.

Einstein, A. "A comment on a Criticism of Unified Field Theory." *PhR*, 89 (1953), p. 321.

Einstein, A. *Mein Weltbild*. Neue, vom Verfasser durchgesehene und wesentlich erweiterte Auflage. Carl Seelig ed. Zürich: Europa, 1953.

Einstein, A. "Letter to C. Seelig, 1953," in: A. Einstein, *Mein Weltbild*. C. Seelig, ed. mit Anmerkungen des Herausgebers (pp. 174-200). Frankfurt/M, Berlin, Wien: Verlag Ullstein GmbH, 1983, p. 199.

Einstein, A. *Foreword*, in: Jammer M., *Concepts of Space. The History of Theories of Space in Physics*. Cambridge, Mass.: Harvard University Press, 1954, pp. 13-14.

Einstein, A. "Fundamental ideas and problems of the theory of relativity." in: *Nobel Lectures*. Published for the Nobel Foundation. Amsterdam-London-New York: Elsevier Publishing Company, 1967, pp. 483-492.

Einstein, A., and B. Kaufman. "Algebraic Properties of the field in the relativistic theory of the asymetric field." *AoM*, 59 (1954), pp. 230-244.

Einstein, A. "Relativity and the Problem of Space (1954)." in: A. Einstein, *Ideas and Opinions*. New York: Crown Publishers, Inc., 5th printing 1960, pp. 360-377.

Einstein, A. "Asymmetric Field Theory." As Appendix II to *The Meaning of Relativity*. 5th edition. Princeton: Princeton University Press, 1955.

Einstein, A. *Prefazione*, in: *Cinquant'anni di relatività (1905-1955)*. Firenze: Giunti Barbèra, 1955, p. 16.

Foster, J., and J. D. Nightingale. *A Short Course in General Relativity*. London: Longmans, 1979.

Frank, Ph. *Einstein. Sein Leben und seine Zeit*. Wiesbaden: F. Vieweg und Sohn, 1979.

Gehrcke E., "Zur Kritik und Geschichte der neueren Gravitationstheorien." *AdP*, 50 (1916), pp. 119-124.

Gehrcke, E. "Über den Äther." *VDPG*, 20 (1918), pp. 165-169.

Gehrcke, E. "Die Relativitätstheorie eine wissenschaftliche Massensuggestion." *Schriften aus dem Verlage der Arbeitsgemeinschaft deutscher Naturforscher zur Erhaltung reiner Wissenschaft e.V.* H. 1. Berlin: 1920

Gehrcke, E. [review] "Ph. Lenard, Über Äther und Uräther." 2. verm. Aufl., *ZftP*, 4 (1923), p. 334.

Gehrcke, E. "Die Gegensätze zwischen der Äthertheorie und Relativitätstheorie und ihre experimentale Prüfung." *ZftP*, 4 (1923), pp. 292-299.

Gerber, P. "Über die räumliche und zeitliche Ausbreitung der Gravitation." *ZMPh*, 43 (1898), p. 93.

Grebe, L. "Über die Gravitationsverschiebung der Fraunhoferschen Linien." *PhZ*, 21 (1920), pp. 662-666.

Guerlac, H. "Newton's Optical Aether." *NRRSL*, 22 (1967), pp. 45-57.

Haves, J.L. "Newton's Revival of the Aether Hypothesis and the Explanation of Gravitational Attraction." *NRRSL*, 23 (1968), pp. 200-212.

Heller, M., "Space-time Structures." *Acta Cosmologica*, 6 (1977), pp. 109-128.

Heller, M. "The Manifold Model for Space-time." *Acta Cosmologica*, 10 (1981), pp. 31-51.

Heller, M. "Relativistic Model for Space-time." *Acta Cosmologica*, 10 (1981), pp. 53-69.

Heller, M. "Time and Causality in General Relativity." *The Astronomy Quarterly*, Vol. 7 (1990), pp. 65-86.

Heller, M. "Jak Einstein stworzył ogólną teorię względności? (How did Einstein create the General Relativity Theory?) *PF*, 39 (1988), pp. 3-21.

Heller, M. *Teoretyczne podstawy kosmologii* (The Theoretical Foundations of Cosmology). Warszawa: PWN, 1988.

Hirosige, T. "Theory of Relativity and the Ether." *Japanese Studies in the History of Science*, 7 (1968), pp. 37-53.

Holton, G. "Mach, Einstein und die wissenschaftliche Suche nach Realität." in: G. Holton, *Thematische Analyse der Wissenschaft. Die Physik Einsteins und seine Zeit*. Frankfurt a. M.: Suhrkamp Verlag, 1981, pp. 203-254.

Howard, D. "Realism and Conventionalism in Einstein's Philosophy of Science: The Einstein-Schlick Correspondence." *Philosophia Naturalis*, 21 (1984), p. 628.

Illy, J. "Mach, Lorentz and Einstein on Ether." in: *Ernst Mach and the Development of Physics*. Intern. Conf., Sept. 14-16, 1988, Prague.

Illy, J. "Einstein Teaches Lorentz, Lorentz Teaches Einstein. Their Collaboration in General Relativity, 1913-1920." *AHESc*, 39 (1989), pp. 247-289.

Infeld, L. "Light Waves in the Theory of Relativity." *Prace Matem. Fiz.*, 32 (1922), pp. 33-84.

Infeld, L., *Szkice z przeszłości*. Warszawa: Państwowy Instytut Wydawniczy, 1964

Israel, H., E. Ruckhaber, and R. Weinmann. *Hundert Authoren gegen Einstein*. Leipzig: R. Voigtlanders Verlag, 1931.

Jammer, M. *Concepts of Space. The History of Theories of Space in Physics*. Cambridge, Mass.: Harvard University Press, 1954.

Jammer, M. "John Stewart Bell and the Debate on Significance of his Contributions to the Foundations of Quantum Mechanics." in: *Bell's Theorem and the Foundations of*

*Modern Physics.* A. Van der Merwe, F. Selleri, G. Tarozzi, eds. Singapore/New Jersey/London/Hong Kong: World Scientific, 1992, pp. 1-23.

K. J., "Der Kampf gegen Einstein." *Die Vossische Zeitung.* in: Paul Weyland. "Betrachtungen über Einsteins Relativitätstheorie und die Art ihrer Einführung." *Schriften aus dem Verlage der Arbeitsgemeinschaft deutscher Naturforscher zur Erhaltung reiner Wissenschafte. V.* H. 2. Berlin: 1920, p. 6.

Klein, M. J. *Paul Ehrenfest.* Vol. 1: *The Making of a Theoretical Physicist.* Amsterdam: North-Holland, 1972.

Kleinert, A. and Ch. Schönbeck. "Lenard und Einstein. Ihr Briefwechsel und ihr Verhältnis vor der Nauheimer Diskussion." *Gesnerus,* 35 (1978), pp. 318-333.

Kostro, L. "Einstein's conception of the ether and its up-to-date applications in the relativistic wave mechanics." in: *Quantum Uncertainties.* W. M. Honig, D. W. Kraft, E. Panarella, eds. New York: Plenum Press, 1987, pp. 435-449

Kostro, L. "Einstein and the ether." *Electronics and Wireless World,* 94 (1988), No. 1625, pp. 238-239.

Kostro, L. "Einstein's new conception of the ether." in: *Physical Interpretations of Relativity Theory.* M. C. Duffy, ed. London: Imperial College, 1988, pp. 55-64.

Kostro, L. *Einstein e l'etere.* Bologna: Società Editrice Andromeda, 1989.

Kostro, L. "Outline of the history of Einstein's relativistic ether concept." in: *Studies in the History of General Relativity.* Einstein Studies. 3. J. Eisenstadt and D. Howard, eds. Boston-Basel-Berlin: Birkhäuser, 1992, pp. 260-280.

Kostro, L. "The Evolution of Einstein's Ideas Concerning Ether, Space and Time." in: *History of Physics in Europe in 19th and 20th Centuries.* F. Bevilacqua Ed. Como, 2-3 Settembre 1992. Bologna: SIF Conference Proceedings, Vol. 42, 1993, pp. 177-183.

Kostro, L. "Albert Einstein and the Theory of the Ether." in: *Foundations of Mathematics & Physics.* eds. U. Bartocci and J. P. Wesley. Blumberg: Benjamin Wesley Publ., 1990, pp. 137-162.

Kostro, L. "The Physical Meaning of Albert Einstein's Relativistic Ether Concept." in: *Frontiers of Fundamental Physics.* M. Barone and F. Selleri, eds.New York-London-Boston: Plenum Publ. Co., 1994, pp. 193-201.

Kox, A. J. "Hendrik Antoon Lorentz, the Ether, and the General Theory of Relativity." *AHESc,* 38 (1988), pp. 67-78.

Körber, H. G. "Zur Biographie des jungen Albert Einstein." *Forschungen und Fortschritte,* 38 (1964), pp. 75-86.

Kuhn, E. "Zweite Weltkraftkonferenz Berlin 1930." *Dinglers politechnisches Journal,* 345 (1930), H. 7, pp. 121-123.

Landau, L. and E. Lifshitz. *Theoria pola.* Warszawa: PWN, 1979.

Laub J. J. "Über die experimentalen Grundlagen des Relativitätsprinzip." *JR,* 7 (1910), pp. 405-420

Laue, M. von. "Zur Erörterung über die Relativitätstheorie." *Tägliche Rundschau* 11/8/1920. Evening edition.

Lenard, Ph. "Über die Lichtemission der Alkaldämpfe und Salze und über die Zentren dieser Emission." *AdP,* 17 (1905).

Lenard, Ph. *Über Äther und Materie.* Zweite, Ausführlichere und mit Zusätzen versehene Auflage. Heidelberg: Carl Winters Universitätsbuchhandlung, 1911.

Lenard, Ph. "Letter to A. Sommerfeld, 4/9/1913." in: A. Kleinert, Ch. Schönbeck, "Ph. Lenard und A. Einstein. Ihr Briefwechsel und ihr Verhältnis vor der Nauheimer Diskussion." *Gesnerus,* 35 (1978), pp. 318-333 (see p. 322).

Lenard, Ph. "Über Relativitätsprinzip, Äther, Gravitation." *JR,* 15 (1918), pp. 117-136.

Lenard, Ph. *Über Relativitätsprinzip, Äther, Gravitation.* Leipzig: Verlag von S. Hirzel, 1918, and later editions of 1920, 1921.

Lenard, Ph. "Zusatz betreffend die Nauheimer Diskussion über das Relativitätsprinzip." in: Ph. Lenard, *Über Relativitätsprinzip, Äther, Gravitation.* Leipzig: Verlag von S. Hirzel, 1921, pp. 36-44.

Lenard, Ph. *Über Äther und Uräther.* Leipzig: Verlag von S. Hirzel, 1921.

Lenard, Ph. *Über Äther und Uräther.* 2. verm. Aufl. Leipzig: Verlag von S. Hirzel, 1922.

Lenard, Ph. "Mahnwort zu Deutsche Naturforscher." in: Ph. Lenard, *Über Äther und Uräther.* 2. verm. Aufl. Leipzig: Verlag von S. Hirzel, 1922, pp. 5-10.

Lenard, Ph. *Grosse Naturforscher.* München: J.F. Lehmanns Verlag, 1929.

Lenard, Ph. *Über Energie und Gravitation.* Berlin and Leipzig: Walter de Gruyter und Co., 1929.

Lorentz, H. A., *Versuch einer Theorie elektrischen und optischen Erscheinungen in bewegten Körpern.* Leiden: Brill, 1895.

Lorentz, H. A. "Der Interferenzversuch Michelsons." in: H.A. Lorentz, *Versuch einer Theorie elektrischen und optischen Erscheinungen in bewegten Körpern.* Leiden: Brill, 1895. pp. 89-92.

Lorentz H. A., "La gravitation." *Scientia,* 16 (1914), pp. 28-59.

Lorentz, H. A. "Letter to A. Einstein, 6/6/1916." *EA* 16-451.

Mach, E. *Die Mechanik in ihrer Entwicklung historisch-kritisch dargestellt.* Leipzig: Brockhaus, 1883.

Mach, E. *Die Mechanik in ihrer Entwiklung historisch-kritisch dargestellt.* Leipzig: Brockhaus, 4. Aufl. 1901, 8 Aufl. 1921.

Mach, E., *The Science of Mechanics, a Critical and Historical Account of its Development.* La Salle (Il.)/London: Open Court, 1942.

Mach, E. *Die Prinzipien der physikalischen Optik.* Leipzig: J.A. Barth, 1921.

Mehra, J. "Albert Einsteins erste Wissenschafliche Arbeit." *PhB,* 27 (1971), pp. 385-389.

Mehra, J. *Einstein, Hilbert, and the Theory of Gravitation.* Dordrecht/Boston: Reidel, 1974.

Miller, A. *Imagery in Scientific Thought Creating 20$^{th}$ Century Physics.* Boston/Stuttgart: Birkhäuser, 1984.

Minkowski, H. "Das Relativitätsprinzip." *AdP,* 47 (1907), pp. 927-938.

Minkowski, H. "Die Grundgleichungen für die elektromagnetischen Vorgänge in bewegter Körper." *Gött. Nachrichten,* 53 (1908).

Minkowski, H. "Raum und Zeit." *PhZ,* 10 (1909), pp. 104-111.

Muller, R. A., "La radiazione cosmica di fondo e la nuova deriva dell'etere." in: *Relatività e cosmologia.* edited by T. Regge. Milano: Le Scienze S.p. A. editore, 1981, pp. 74-84

Newton, I. *Mathematical Principles.* Cambridge: Cambridge Univ. Press, 1934.

Norton, J. "How Einstein found his field equations?" in: *Einstein and the History of General Relativity.* D. Howard, J. Stachel, eds. Einstein Studies, Vol. 1. Boston/Basel/Berlin: Birkhauser,1989. pp. 101-159.

Ostwald, W. *Lehrbuch der allgemeinen Chemie.* (In zwei Banden) Leipzig: W. Engelmann, 1893.

Pais, A. *"Subtle is the Lord..." The Science and the Life of Albert Einstein.* New York/Oxford: Clarendon Press/Oxford University Press, 1982.

Penrose, R. "The Mass of the Classical Vacuum." in: S. Saunders and H. R. Brown, eds. *The Philosophy of Vacuum.* Oxford: Clarendon Press, 1991, pp. 21-26.

Piaggio, H.T.H. "The Concept of Space." (elaboration of A. Einstein's Address to the University of Nottingham), *Nature,* 125 (1930), pp. 897-898.

Poincaré, H. *La science et l'hypothèse.* Paris: Flammarion, 1968.

Prokhovnik, S. J. "A Cosmological Basis for Bell's View on Quantum and Relativistic Physics." in: *Bell's Theorem and the Foundation of Modern Physics.* A Van Merwe, F. Selleri and G. Tarozzi, eds. New Jersey/London: World Scientific, 1992 p.388.

Prokhovnik, S. J. *Light in Einstein's Universe.* Dordrecht: Reidel, 1985

Rosenfeld, L. "Newton's Views on Aether and Gravitation." *AHESc,* 6 (1969), pp. 29-37.

Saunders, S., and H. R. Brown, eds. *The Philosophy of Vacuum.* Oxford: Clarendon Press, 1991.

Saunders, S., and H. R. Brown. "Reflections on Ether." in: S. Saunders and H. R. Brown, eds. *The Philosophy of Vacuum.* Oxford: Clarendon Press, 1991, pp.27-63.

Schrempf, Ch. *Der Weltäther als Grundlage eines einheitlichen Weltbildes.* Leipzig: Otto Hillmanns Verlag, 1934.

Selleri, F. "Space-time Transformations in Ether Theories." *Z. Naturforsch.* 46a (1990), p. 419-425.

Selleri, F. "Special Relativity as a Limit of Ether Theories." in: *Physical Interpretations of Relativity Theory.* London: British Society for Philosophy of Science, 1990 pp. 508-514.

Selleri, F., "Theories Equivalent to Special Relativity." in: *Frontiers of Fundamental Physics.* M. Barone and F. Selleri eds. New York: Plenum, 1994.

Selleri, F. "Inertial Systems and the Transformations of Space and Time." *Physics Essays,* 8, No 3 (1995), pp.342-349

Selleri, F. *Found. Phys.* 126 (1996), p.641.

Selleri, F., ed. *Open Questions in Relativistic Physics.* Proceedings of the International Conference, Athens, Greece, 25-28 June 1997. Montreal: Apeiron, 1998.

Selleri, F. "On the Existence of a Physical and Mathematical Discontinuity in Relativistic Theory." in: Selleri, F.ed., *Open Questions in Relativistic Physics.* Proceedings of the International Conference, Athens, Greece, 25-28 June 1997. Montreal: Apeiron, 1998.

Solovine, M. "Introduction..." in: A. Einstein, *Briefe an Maurice Solovine (1906-1955).* Berlin: Veb deutscher Verlag der Wissenschaften, 1960, pp. VII-XV.

Średniawa, B. " The reception of the Theory of Relativity in Poland." in: Thomas F. Glick (ed.), *The Comparative Reception of Relativity*, D. Reidel Publishing Company, Dordrecht/ Boston/Lancaster/Tokyo 1987, pp. 327-350.

Średniawa, B. "The Evolution of the Concept of Ether and the Early Development of Relativity at Cracow University." in: *Universitas Iagellonica Acta Scientiarum Litterarumque*, MCLI, Universitatis Iagellonicae Folia Physica, Fasciculus XXXVII, Kraków 1994, pp. 9-20.

Średniawa, B. "Early Investigations in the Foundations of General Relativity." in: *Universitas Iagellonica Acta Scientiarum Litterarumque*, MCLI, Universitatis Iagellonicae Folia Physica, Fasciculus XXXVII, Kraków 1994, pp. 21-37.

Średniawa, B. "Three Essays on History of Relativity in Cracow." in: *Universitas Iagellonica Acta Scientiarum Litterarumque*, MCLI, Universitatis Iagellonicae Folia Physica, Fasciculus XXXVII, Kraków 1994.

Średniawa, B. "Kontakty naukowe i współpraca polskich fizyków z Einsteinem." *Kwartalnik Historii Nauki i Techniki*, Vol. 41 (1996), No 1 p. 59-97.

Stachel, J. "Einstein's Search for General Covariance, 1912-1915." in: *Einstein and the History of General Relativity*. Boston-Basel-Berlin: Birkhäuser, 1989.

Stachel, J., and D. Howard, eds. *Einstein and the History of General Relativity*. Boston-Basel-Berlin: Birkhäuser, 1989.

Stark, J. *Die gegenwärtige Krisis in der deutschen Physik*. Leipzig: Verlag von J.A. Barth, 1922.

Tatarkiewicz, W. *Historia filozofii*. 5th ed., Vol. 3. Warszawa: PWN, 1958.

Vigier, J.-P. "New non-zero photon mass interpretation of Sagnac effect as direct experimental justification of the Langevin paradox." *Phys. Lett. A*, 234 (1997), pp. 75-85.

Vizgin, V. P. and Ya. A. Smorodinskii. "From the equivalence principle to the equation of gravitation." *Sov. Phys. Usp.*, 22 (7), July 1979, pp. 489-515.

Wahsner, R. "Äther und Materie–von Descartes bis Einstein." *Wissenschaft und Fortschritte* 29 (1979), pp. 54-57.

Weyl, H. *Raum, Zeit, Materie*. Berlin: Verlag von J. Springer, 1918.

Weyl, H. "Elektrizität und Gravitation." *PhZ*, 21 (1920), pp. 649-651.

Weyl, H. "Feld und Materie." *AdP*, 65 (1921), pp. 541-563.

Weyl, H. "Die Relativitätstheorie auf der Naturforscherversammlung in Bad Nauheim." *Jahresbericht der Deutschen Mathematikervereinigung* 31 (1922), pp. 51-63.

Weyland, P. "Einsteins Relativitätstheorie–eine wissenschaftliche Massensuggestion." *Tägliche Rundschau*, Freitag 6/8/1920, Abendausgabe.

Weyland, P. *Tägliche Rundschau*, Mittwoch 11/8/1920, Abendausgabe.

Weyland, P. "Betrachtungen über Einsteins Relativitätstheorie und die Art ihrer Einführung." *Schriften aus dem Verlage der Arbeitsgemeinschaft deutscher Naturforscher zur Erhaltung reiner Wissenschaft* e. V. H. 2., Berlin 1920, pp. 10-20.

Whittaker, E. *A History of the Theories of Aether and Electricity*. Vol. 1-2. New York: Harper and Brothers, 1953.

# Index of Proper Names

Made in the USA
Middletown, DE
28 July 2017